M. Sandra Wood
Janet M. Coggan
Editors

Men's Health on the Internet

Men's Health on the Internet has been co-published simultaneously as *Health Care on the Internet*, Volume 6, Numbers 1/2 2002.

Pre-publication REVIEWS, COMMENTARIES, EVALUATIONS . . .

"VALUABLE for anyone searching for a compilation of reliable Web sites. Readers of this well-organized resource will be armed with good, basic Web sites that will provide an important foundation for self-learning."

Roberta Bronson Fitzpatrick, MLIS
Associate Director
George F. Smith Library of the Health Sciences
University of Medicine & Dentistry of New Jersey

More pre-publication
REVIEWS, COMMENTARIES, EVALUATIONS . . .

"MAGNIFICENT. . . . The authors have compiled only the best resources. . . . With so much information now available on the Internet, this book provides an invaluable service to help readers pinpoint reliable health information."

Jana C. Allcock, MLS, AHIP
Consumer Health Outreach Coordinator
National Network of Libraries of Medicine
Southeastern/Atlantic Region

"EXCELLENT. . . . Given the propensity for the males of the human species to be far less proactive in their own health care than females, this book provides an excellent jumping-off point for men. The authors have done an excellent job. . . . Covers everything from impotence and prostate cancer to hair loss and male menopause. Using this book, men will be able to 'cut to the chase' in the morass that is the Internet."

Thomas L. Williams, MSLS, AHIP
Director
C. M. Baugh Biomedical Library
and Media Production Services
University of South Alabama

More pre-publication
REVIEWS, COMMENTARIES, EVALUATIONS . . .

"THE AUTHORS HAVE PROVIDED A REAL SERVICE . . . A HIGHLY SELECTIVE LIST OF RESOURCES. . . . Most popular doesn't necessarily mean the most reliable or quality-based. The introductory information for each area covered is written using lay terminology, but substantive enough to be valuable to all levels of readers."

J. Michael Homan, MA
Director of Libraries
Mayo Clinic

"An expertly selected, well-annotated set of Web sites. . . . The authors do a superb job of cutting through the chaff to identify sites that present well-organized, authoritative information. When one looks at the thousands of hits an Internet search provides for a typical men's health topic, the necessity of a book like this is readily apparent. AN EXCELLENT RESOURCE, particularly for public libraries and consumer health collections."

T. Scott Plutchak, MA
Editor
Journal of the Medical Library Association

The Haworth Information Press
An Imprint of The Haworth Press, Inc

Men's Health on the Internet

Men's Health on the Internet has been co-published simultaneously as *Health Care on the Internet*, Volume 6, Numbers 1/2 2002.

Health Care on the Internet Monographic "Separates"

Below is a list of "separates," which in serials librarianship means a special issue simultaneously published as a special journal issue or double-issue *and* as a "separate" hardbound monograph. (This is a format which we also call a "DocuSerial.")

"Separates" are published because specialized libraries or professionals may wish to purchase a specific thematic issue by itself in a format which can be separately cataloged and shelved, as opposed to purchasing the journal on an on-going basis. Faculty members may also more easily consider a "separate" for classroom adoption.

"Separates" are carefully classified separately with the major book jobbers so that the journal tie-in can be noted on new book order slips to avoid duplicate purchasing.

You may wish to visit Haworth's website at . . .

http://www.HaworthPress.com

. . . to search our online catalog for complete tables of contents of these separates and related publications.

You may also call 1-800-HAWORTH (outside US/Canada: 607-722-5857), or Fax 1-800-895-0582 (outside US/Canada: 607-771-0012), or e-mail at:

getinfo@haworthpressinc.com

Men's Health on the Internet, edited by M. Sandra Wood, MLS, MBA, and Janet M. Coggan, MEd, MSLS (Vol. 6, No. 1/2, 2002). *Shows how the Internet can be used to locate accurate information about the diagnosis, treatment, and prognosis of men's health problems.*

Women's Health on the Internet, edited by M. Sandra Wood, MLS, MBA, and Janet M. Coggan, MEd, MSLS (Vol. 4, No. 2/3, 2000). *A guide to Web sites representing women's health organizations, sites where you can find current women's health news, and much more.*

HIV/AIDS Internet Information Sources and Resources, edited by Jeffrey T. Huber, PhD (Vol. 2, No. 2/3, 1998). *"Will demystify and detangle the World Wide Web for the AIDS researcher, the primary care provider, and the patient anxious for first-hand information in various levels of detail." (Kiyoshi Kuromiya, Director, Critical Path AIDS Project, Philadelphia, Pennsylvania)*

Cancer Resources on the Internet, edited by M. Sandra Wood, MLS, MBA, and Eric P. Delozier, MLS (Vol. 1, No. 3, 1997). *"A refreshing and welcome emphasis on what's on the Internet for medicine consumers together with a wealth of useful resources and techniques." (Alan M. Rees, Professor Emeritus, Case Western Reserve University, Editor, Consumer Health and Nutrition Index, Author, Consumer Health Information Source Book, 4th Edition)*

Mens's Health on the Internet

M. Sandra Wood, MLS, MBA
Janet M. Coggan, MEd, MSLS
Editors

Men's Health on the Internet has been co-published simultaneously as *Health Care on the Internet,* Volume 6, Numbers 1/2 2002.

The Haworth Information Press
An Imprint of
The Haworth Press, Inc.
New York • London • Oxford

Published by

The Haworth Information Press®, 10 Alice Street, Binghamton, NY 13904-1580 USA

The Haworth Information Press® is an imprint of The Haworth Press, Inc., 10 Alice Street, Binghamton, NY 13904-1580 USA.

Men's Health on the Internet has been co-published simultaneously as *Health Care on the Internet*, Volume 6, Numbers 1/2 2002.

Cover design by Jennifer Gaska.

Library of Congress Cataloging-in-Publication Data

Wood, M. Sandra
Men's health on the Internet / M. Sandra Wood.
p. cm.
Includes bibliographical references and index.
ISBN 0-7890-1924-8 (hard: alk. paper)–ISBN 0-7890-1925-6 (pbk.: alk. paper)
1. Men–Health and hygiene. 2. Internet. 3. Health–Computer network resources. 4. Self-care, Health. I. Title.

RA777.8.W66 2002
025.06'613'04234–dc21

2002017208

Indexing, Abstracting & Website/Internet Coverage

This section provides you with a list of major indexing & abstracting services. That is to say, each service began covering this periodical during the year noted in the right column. Most Websites which are listed below have indicated that they will either post, disseminate, compile, archive, cite or alter their own Website users with research-based content from this work. (This list is as current as the copyright date of this publication.)

Abstracting, Website/Indexing Coverage . . . . Year When Coverage Began

- *Adis International Ltd.* 1998
- *AgeLine Database* 1997
- *Applied Social Sciences Index & Abstracts (ASSIA) (Online: ASSI via Data-Star) (CDRom: ASSIA Plus) <www.csa.com>* 1997
- *BUBL Information Service: An Internet-based Information Service for the UK higher education community <URL:http://bubl.ac.uk/>* 1997
- *Cambridge Scientific Abstracts (Web Resources Database) <www.csa.com>* 1997
- *CINAHL (Cumulative Index to Nursing & Allied Health Literature), in print, EBSCO, and SilverPlatter, Data-Star, and Paper Chase. (Support materials include Subject Heading List, Database Search Guide, and instructional video.) <www.cinahl.com>* 1997
- *CNPIEC Reference Guide: Chinese National Directory of Foreign Periodicals* 1997
- *Combined Health Information Database (CHID)* 1997
- *Computing Reviews* 1997

(continued)

- *Current Awareness Abstracts of Library & Information Management Literature, ASLIB (UK)* 1997
- *e-psyche, LLC <www.e-psyche.net>* 2001
- *European Association for Health Information & Libraries: selected abstracts in newsletter "Publications" section* 1997
- *FINDEX <www.publist.com>* 1999
- *Health Care Literature Information Network/HECLINET* 1997
- *Health Service Abstracts (HSA)* 1997
- *Healthcare Leadership Review* 1997
- *Index Guide to College Journals (core list compiled by integrating 48 indexes frequently used to support undergraduate programs in small to medium sized libraries)* 1999
- *Index to Periodical Articles Related to Law* 1997
- *Information Science Abstracts <www.infotoday.com>* 1997
- *Internet & Personal Computing Abstracts <www.infotoday.com/mca/default.htm>* 1999
- *INSPEC <www.iee.org.uk/publish/>* 1997
- *Leeds Medical Information* 1997
- *Library Association Health Libraries Group Newsletter <www.blackwell-science.com/hlr/newsletter/>* 1998
- *Library & Information Science Abstracts (LISA) <www.csa.com>* . . 1997
- *Library and Information Science Annual (LISCA) <www.lu.com/arba>* 1999
- *MANTIS (Manual, Alternative and Natural Therapy) <www.healthindex.com>* 2000
- *Medicinal & Aromatic Plants Abstracts (MAPA)* 1997
- *PASCAL, c/o Institute de L'Information Scientifique et Technique <http://www.inist.fr>* 1997
- *Patient Care Management Abstracts* 1997

(continued)

- ***Pharmacy Business*** **. 1997**
- ***Referativnyi Zhurnal (Abstracts Journal of the All-Russian Institute of Scientific and Technical Information–in Russian)*** **. . . 1997**
- ***Social Work Abstracts <www.silverplatter.com/catalog/swab.htm>*** **. . 1997**

Special Bibliographic Notes related to special journal issues (separates) and indexing/abstracting:

- indexing/abstracting services in this list will also cover material in any "separate" that is co-published simultaneously with Haworth's special thematic journal issue or DocuSerial. Indexing/abstracting usually covers material at the article/chapter level.
- monographic co-editions are intended for either non-subscribers or libraries which intend to purchase a second copy for their circulating collections.
- monographic co-editions are reported to all jobbers/wholesalers/approval plans. The source journal is listed as the "series" to assist the prevention of duplicate purchasing in the same manner utilized for books-in-series.
- to facilitate user/access services all indexing/abstracting services are encouraged to utilize the co-indexing entry note indicated at the bottom of the first page of each article/chapter/contribution.
- this is intended to assist a library user of any reference tool (whether print, electronic, online, or CD-ROM) to locate the monographic version if the library has purchased this version but not a subscription to the source journal.
- individual articles/chapters in any Haworth publication are also available through the Haworth Document Delivery Service (HDDS).

ALL HAWORTH INFORMATION PRESS
BOOKS AND JOURNALS ARE PRINTED
ON CERTIFIED ACID-FREE PAPER

Men's Health on the Internet

CONTENTS

ABOUT THE EDITORS

M. Sandra Wood, MLS, MBA, is Librarian, Reference and Database Services, of The Milton S. Hershey Medical Center at The Pennsylvania State University in Hershey. She holds the academic rank of librarian and has over thirty-one years of experience as a medical reference librarian, including the areas of general reference services, management of reference services, database and Internet searching, and user instruction. Ms. Wood has been widely published in the field of medical reference and is Editor of the journals *Medical Reference Services Quarterly* and *Health Care on the Internet* (The Haworth Press, Inc.) and of several Haworth books, including *Women's Health on the Internet*, *Health Care Resources on the Internet: A Guide for Librarians and Health Care Consumers*, and *Cancer Resources on the Internet*. She is a member of the Medical Library Association and the Special Libraries Association, and has served on the MLA's Board of Directors as Treasurer. Ms. Wood is also a Fellow of the Medical Library Association.

Janet M. Coggan, MEd, MSLS, worked as the Area Health Education Center Coordinator and as Reference Librarian in the Reference Department of the University of Florida Health Science Center (HSC) Library in Gainesville, Florida for nine years until fall 1998. During that time, Ms. Coggan was very active in the Consumer and Patient Health Information Section of the Medical Library Association and developed a consumer health collection for the HSC Library. In addition, she has served as the editor of the book review column for *Medical Reference Services Quarterly* for three years and currently serves as the book review editor for the *Journal of the Medical Library Association*. Ms. Coggan has published numerous book reviews in those journals as well as in *Choice* and *Library Journal*. In addition, she was co-editor of *Women's Health on the Internet* and wrote the chapter, "Consumer Health Information on the Internet," for the recent Haworth Press publication *Health Care Resources on the Internet: A Guide for Librarians and Health Care Consumers*. She also writes content for the Internet site DocsDepot.org. Currently working as a consultant, Ms. Coggan is based in Gainesville, Florida.

Finding Information About Men's Health on the Internet: An Overview

M. Sandra Wood

SUMMARY. It's a well-known fact that men are less likely than women to seek health care. Men tend to wait as long as possible to see a doctor about a health problem, making early diagnosis of a disease difficult if not impossible. When one looks at the statistics related to men's health, the apparent aversion to seeking health care is disconcerting. Given the fact that men seem to be more private about their illnesses, the Internet provides an excellent place for them to seek health care information. A search of "men's health" finds an overwhelming number of sites on the Internet. This article provides a highly selected list of Web sites intended as a starting place for finding information about men's health on the Internet. *[Article copies available for a fee from The Haworth Document Delivery Service: 1-800-HAWORTH. E-mail address: <getinfo@haworthpressinc.com> Website: <http://www.HaworthPress.com> © 2002 by The Haworth Press, Inc. All rights reserved.]*

KEYWORDS. Men's health, Internet

INTRODUCTION

It's a well-known fact that men are less likely than women to seek health care. In fact, "women visit the doctor 150 percent as often as

---

M. Sandra Wood, MLS, MBA (mswood@psu.edu) is Librarian, Reference and Database Services, The George T. Harrell Library, The Milton S. Hershey Medical Center, The Pennsylvania State University, P.O. Box 850, Hershey, PA 17033-0850.

[Haworth co-indexing entry note]: "Finding Information About Men's Health on the Internet: An Overview." Wood, M. Sandra. Co-published simultaneously in *Health Care on the Internet* (The Haworth Information Press, an imprint of The Haworth Press, Inc.) Vol. 6, No. 1/2, 2002, pp. 1-16; and: *Men's Health on the Internet* (ed: M. Sandra Wood, and Janet M. Coggan) The Haworth Information Press, an imprint of The Haworth Press, Inc., 2002, pp. 1-16. Single or multiple copies of this article are available for a fee from The Haworth Document Delivery Service [1-800-HAWORTH, 9:00 a.m. - 5:00 p.m. (EST). E-mail address: getinfo@haworthpressinc.com].

men."[1] "According to a recent survey by *Men's Health* magazine and CNN, one-third of American men have not had a checkup in the past year. Nine million men haven't seen a doctor in five years."[2] Additionally, another survey found that "slightly more than 15 million men have not had a basic health check in five years or more."[1] The implication is that there is a "male denial factor" about visiting a doctor.[1] Men tend to wait as long as possible to see a doctor about a health problem, making early diagnosis of a disease difficult if not impossible.

When one looks at the statistics related to men's health, the apparent aversion to seeking health care is disconcerting. According to another source, by age 75, "men die of cancer at about twice the rate of women" and the death rate for prostate cancer is nearly twice what it is for breast cancer.[3] However, despite the discrepancy between sexes on willingness to see a doctor, the U.S. Center for Disease Control's National Center for Health Statistics indicates that in 1999, 311 million men made an annual office visit to a physician; men visited hospital outpatient departments 33.6 million times and emergency departments 49 million times in 1999; and, there were 16.4 million surgical procedures performed on men in 1999.[4]

Given the fact that men seem to be more private about their illnesses, the Internet provides an excellent place for them to seek health care information. According to a recent study, 52 million American adults have sought health care information on the Internet either for themselves or for family members and friends.[5] This same study indicates that "men are slightly more privacy-conscious . . . And men are somewhat more eager to take advantage of the fact that they feel anonymous online; they are more likely to have used the Web to search for information about sensitive health issues."[5] Convenience and privacy for seeking information are features that will likely increase the Internet's value as a source for health care information.

Nationally in the U.S., there is also a trend for people to take a greater role in the management of their health care. With a greater expectation of participation in health care decisions, a recent study projects the growth of Internet health users to increase from nearly 25 million in 1999 to over 52 million in 2003.[6] This study, performed by Cyber Dialogue, examined the impact of the Internet on doctor selection, doctor-patient roles, and patient education. While this study found a distrust of the Internet on the part of both doctors and patients due to quality of information, most users said "they expect their use to stay the same or increase in the future."[6]

SITE SELECTION

A search of the Internet was conducted in early September 2001 on "men's health" using search engines that health care consumers–patients and/or their families–might use. The results were overwhelming both in the number of sites found and in the inconsistency of the quality of the information. Looksmart listed 916 "reviewed" sites; Google found 720,000 sites; Excite, over 4.5 million sites; and AltaVista, 43 million sites.

An initial glance at the sites found with this basic search revealed everything from personal experiences and sites selling cures for diseases to some very high quality sites offering sound medical information. The sheer volume of information found was overwhelming; there was simply too much information to sort through. Additionally, there was a heavy emphasis on body building and physical appearance/body image, so it was necessary to scroll through and eliminate those sites, which proved time consuming. Clearly, a general search of the Internet was not the most productive way to start.

The problem of finding trusted, quality information on the Internet remains a problem. Unfortunately, for inexperienced Internet users, there often is no easy no way to distinguish the good sites from the bad. Several organizations have developed criteria to evaluate health care Web sites. Perhaps the best known is the Health On the Net Foundation (HON) <http://www.hon.ch/HONcode/Conduct.html>. Sites which subscribe to such codes are generally considered to be more reliable than others. However, this still does not resolve the problem of the overwhelming volume of Web sites.

The following list of Web sites is intended as a starting place for finding information about men's health, thus eliminating the need to perform a "net search" in a general search engine. Most of these sites will also be mentioned elsewhere in this work in the articles that deal with specific aspects of men's health. Sites were selected based on the searching experience and knowledge of the author. Additionally, they all meet general criteria that include unbiased content, good design, disclosure of authors/sponsors, current information, easy access and long-standing availability, and updated quality links. The sites are intended for the general public, rather than the health care professional. In short, this is a highly selective list of sites and is in no way a comprehensive list, even of reviewed material.

The sites are listed alphabetically, with no division by type of site or by special subject area. Most sites are made available by educational institutions (.edu), professional organizations (.org), and government sites (.gov). Several commercial sites (.com) are included because they provided valuable, impartial information. Sites focusing on specific subject areas (e.g., prostate cancer, heart disease) are not included here, but will be found elsewhere in this work. Where the direct path to the men's health page is easy to list, that path is given; where the URL is more complicated, the general site is listed with instructions on how to locate the men's page.

MEN'S HEALTH WEB SITES

Ask NOAH About: Men's Health <http://www.noah-health.org/english/wellness/healthyliving/menshealth.html>

NOAH (New York Online Access to Health) is a site sponsored by public libraries and health agencies in New York City (see Figure 1). This site <http://www.noah-health.org> can be listed as a core information site for almost any health topic because it contains links to information at sites selected and reviewed by volunteer subject editors. Thus, the information is screened and non-biased. Because of the extensive information also available in Spanish, NOAH is the site of choice for information in that language. Adobe Acrobat is required to view the PDF files.

NOAH's Men's Health page is divided into Basic Information, Specific Issues, and Information Resources. Basic Information includes topics such as "Screening Tests for Men" along with a section on male anatomy. Examples of Specific Issues include "Breast Cancer in Men," "Preserving Your Fertility: Risk Factors," "Hair Loss," "Prostate Cancer," "Sexuality," and "Testicular Self Exam." Links take you to professional medical associations and other reputable Web sites. The Information Resources section has links to 3 other major Web sites containing men's health information.

drkoop.com: Men's Health <http://www.drkoop.com/dyncon/center.asp?id=1024>

Drkoop.com is a company led by former U.S. Surgeon General, Dr. C. Everett Koop. The Web site is based on Dr. Koop's vision that "peo-

FIGURE 1. Ask NOAH About: Men's Health Web Page

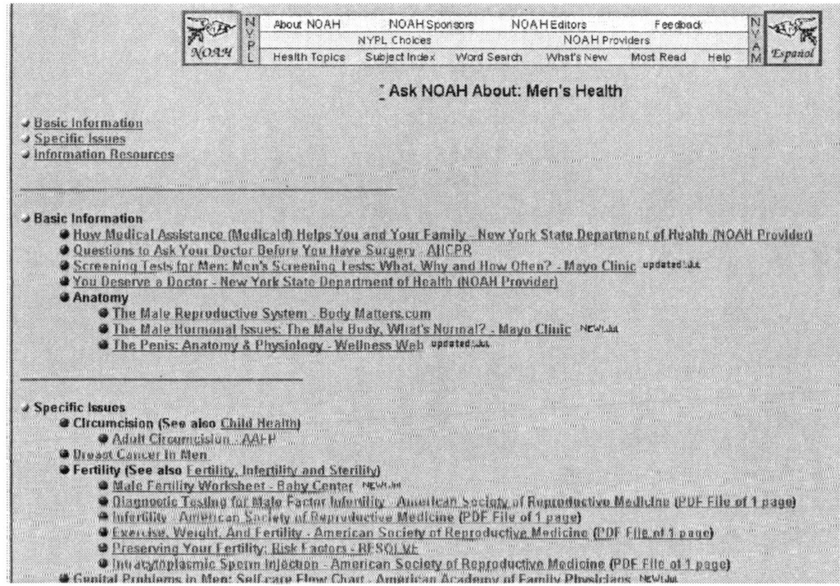

ple should be empowered to better manage their personal health with comprehensive, trusted information."

The "Men's Health" page can be reached either directly from the URL listed above, or by going to <http://www.drkoop.com> and selecting "Men's Health" under either "My Health" or "Family" (see Figure 2). The page has a main story, which is changed frequently (one day it was "Angry Young Men" and another, "Testicular Cancer–One Man's Journey") and eventually ends up in the "Archive," which is found under "Men's Health." These articles, listed in no apparent order (perhaps by date), include topics ranging from heartburn, binge drinking, and low back pain to prostate disorders, male infertility, and stress management.

One note of caution with this site is that is has many commercial sponsors, and a variety of products are available for sale from most pages. Articles with specific bias are labeled as "Sponsored Content."

familydoctor.org
<http://www.familydoctor.org>

This is the patient information site of the American Academy of Family Physicians, a "national, non-profit medical association of 93,100

FIGURE 2. drkoop.com: Men's Health Web Page

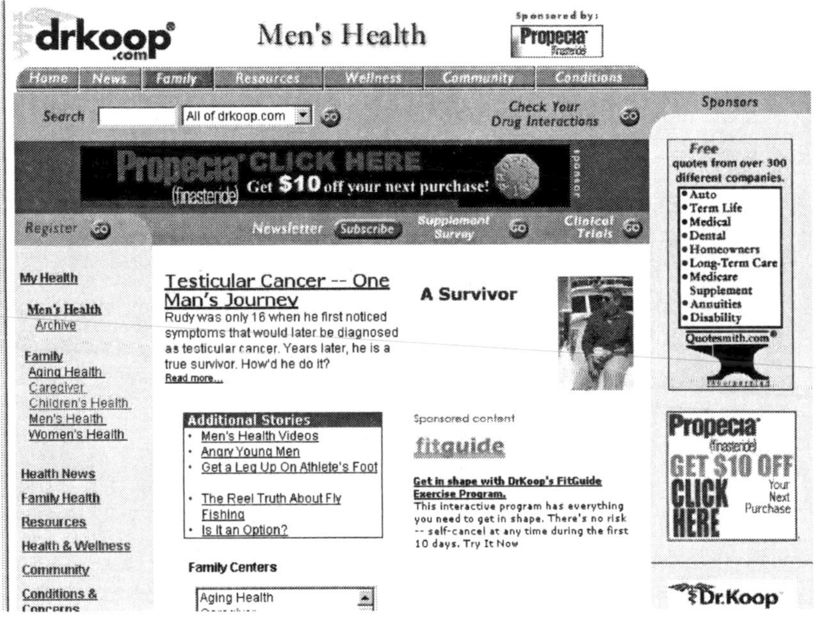

members, including family physicians, family practice residents and medical students." The page advertises that it provides information for the whole family, and has divided its information into four categories: children, women, men, and seniors.

Selecting the "Men" picture on the right side of the page (or searching for "men" in "Special Groups") takes one to an alphabetic listing of topics that are related to men's health (see Figure 3). This simplistic arrangement allows direct access to all relevant topics from this one page. Examples of topics are "Breast Problems in Men," "Gout," "Hair Loss," "Impotence," and "Sleep Apnea," and "Viagra." Topics are codes as "Patient Handout," "Self-Care," "Health Fact," and "Spanish Version Available." The patient handouts include basic, factual information, along with relevant information. Topics coded as "Self Care" are basically decision charts for diagnosis. "Health Facts" are taken from the "AAFP Family Health Facts" series and could be viewed but not printed; they are available as brochures in bulk.

FIGURE 3. familydoctor.org: Men's Health Web Page

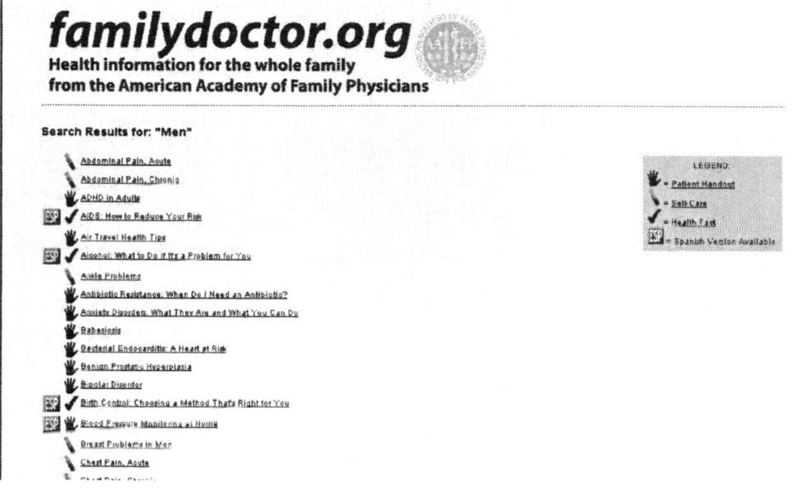

HealthAtoZ: Men's Health
<http:www.healthatoz.com/atoz/centers/menshealth/mensindex.html>

HealthAtoZ is owned by Medical Network, Inc., an Internet health information company. The Men's Health page on HealthAtoZ can be accessed directly via the above URL, or you can go to the main page <http://www.healthatoz.com> and select "men's Health" listed under "Your Family" (see Figure 4). This site subscribes to the HON code.

The "Men's Health" page has sections on "Prevention," "Symptoms & Remedies," "Nutrition & Fitness," and "Smokeout Center." Links to documents are available directly off this page. At the time of this review subjects included immunizations, high blood pressure, baldness, and impotence. The commercial side of this site is exemplified at the end of the impotence document where the source is listed as Viagra.com; the active link takes you to the drug company's site. This site also allows posting of messages on its "Men's Health Message Board."

InteliHealth: Men's Health
<http://www.intelihealth.com>

InteliHealth indicates that its "mission is to empower people with trusted solutions for healthier living . . . " The information provided is

FIGURE 4. Health A to Z: Men's Health Web Page

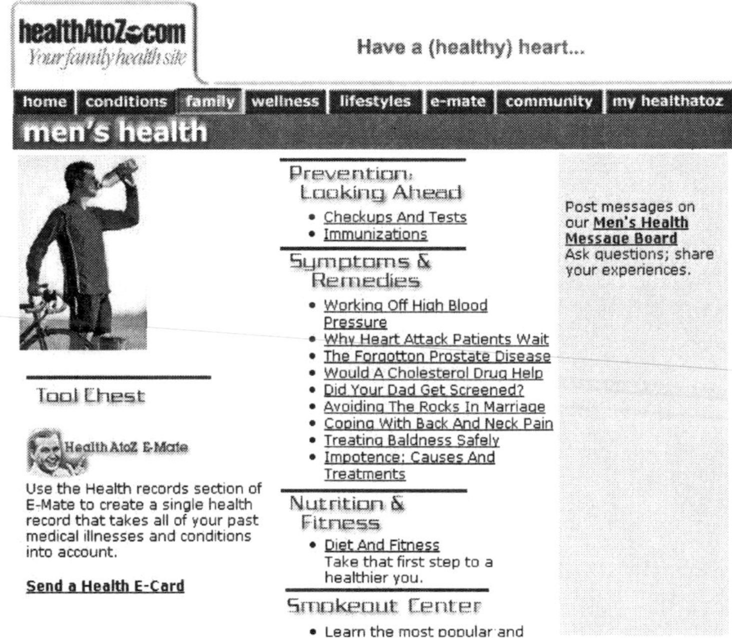

from Harvard Medical School and the University of Pennsylvania School of Dental Medicine, with contributions from other health care organizations, government agencies, and the media. This site is a great place to start looking for health care information in general.

To get to the Men's Health page on InteliHealth, select "Men's Health" from the bar that runs near the top of the page (see Figure 5). This page contains a combination of basic health and news items. On this particular day, the "News" featured stories on puberty, berries and cancer, and low sperm count. Featured topics included the enlarged prostate, sports safety, and Viagra. "Basics" discussed visits to the doctor's office and health screening, while "Just for Guys" had sections on "Your Appearance," "Your Work Life," "Your Sex Life," and "Your Mental Well Being." An "Interactive Tools" section provides an opportunity for online interaction (e.g., "How Many Beats, How Many Breaths").

FIGURE 5. InteliHealth: Men's Health Web Page

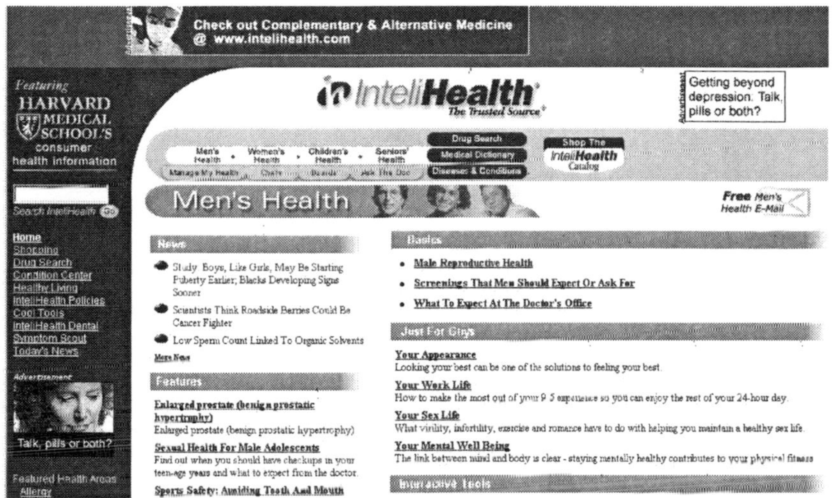

One note of criticism with this site is that the actual information is buried several levels deep. For example, after choosing "Your Appearance," one is taken to the next level, which lists "Hair," "Skin," and "Weight." Selecting "Weight" leads to the next level, which has several choices. Selecting "Evaluating Your Physical Condition" finally leads to an excellent presentation on physical fitness.

InteliHealth accepts advertisements, but the distinction between ads and health information is clear. This site is a founding member of Hi-Ethics and subscribes to the HONCode.

Male Health Center <http://www.malehealthcenter.com>

The Male Health Center, located in Dallas, Texas, "was the first center in the United States specializing in male health." Founded in 1989 by Dr. Kenneth A. Goldberg, the center provided "men with an integrated system of care that addressed all their needs."

The Web site for the Male Health Center presents the center's philosophy of a holistic approach to health care. Many of the articles include multiple treatment methods, rather than singling out one approach. Men are urged to get involved with their medical care. This site is intended as an education site, both for the male patient and his partner (see Figure 6).

FIGURE 6. Male Health Center Web Page

The Site Map divides the information into sections such as "About Male Health," which includes articles about cancer, the prostate, vasectomy; "Symptoms" (e.g., frequent urination, low back pain); "Self Care" (e.g., diet, exercise, screening); "Sex and Other Fun Stuff" (e.g., impotence, sexually transmitted diseases); "Straight Talk" (a section of questions and answers); and "His Health" (columns by Dr. Goldberg). The organization allows for several ways to find the same information; for example, "frequent urination" and "prostate enlargement" both lead to the same article. The material is written in an easy-to-read style, providing an excellent starting place for men's health information.

MayoClinic: Men's Health Center
<http://www.mayohealth.org/home?id=4.1.4>

From the MayoHealth main page <http://www.mayohealth.com>, the Men's Health Center can be reached by selecting "Men's Health" in the pull-down menu under "Healthy Living Centers," or by going directly to the page using the URL listed above. The Mayo Clinic Web page is another resource that can be used to access information on almost any health topic. The material is written for the consumer and comes from a reputable source, making it a great place to begin for anyone just beginning a search on the Internet for health information.

The Men's Health Center brings together, on one page, health issues that are relevant to men (see Figure 7). "News and Feature Picks" contains timely articles (e.g., "Spit Tobacco: Does Smokeless Mean Harmless?") on the main page, with a link to more articles. "Conditions and Care" goes to a page with sections linking to articles on "Cancers," "Drugs," "Heart Disease," "Hormonal Conditions," "Mental Health," "Other," "Sexuality and Reproduction," "Tests and Procedures," and "Urologic Conditions." The section on "Lifestyle" links to sections with articles on "Appearance," "Finding a Balance," "Fitness and Weight," "Relationships," "Risky Business," and "Working Life." "Be-

FIGURE 7 MayoClinic Men's Health Center Web Page

ing Male" links to articles grouped as "Male Hormonal Issues," "Reproduction," and "Screening Tests for Men." At the bottom of the page, "Take Charge of Your Health" includes "planners" on weight, exercise, and stress management.

Medem: Men's Health <http://www.medem.com/MedLb/sub_detaillb.cfm?parent_id=55&ace=disp>

The "Men's Health" page of Medem is accessible via the above URL, or you can go to the general site <http://www.medem.com>, select "Medical Library," and then select "Men's Health" from the "Life Stages" section (see Figure 8). Medem is sponsored by over 25 professional medical societies, including the American Medical Association. In fact, selection of information for patients on the AMA site links users to this site.

The information provided on Medem is excellent and of the highest quality, consisting of peer-reviewed articles. However, the site is lacking in its physical presentation and difficult to navigate. "Men's Health" is divided into six sections, including "Healthy Living" and "Reproductive Health and Sexual Function." Following the link to the next level brings up a list of documents. Because of the type font and organization,

FIGURE 8. Medem Men's Health Web Page

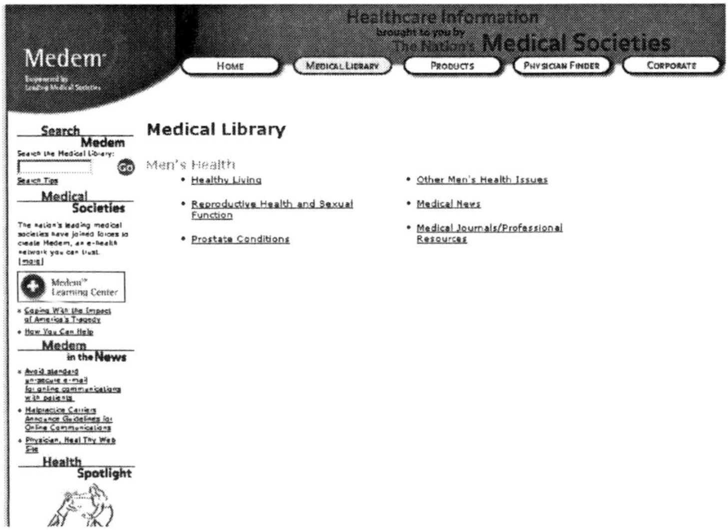

it's difficult to determine your location within the Web site. Once a document is brought up, however, the type font and presentation change, and the content of the document is worth the effort to find it.

Men's Health Network
<http://www.menshealthnetwork.org>

The Men's Health Network "is a non-profit educational organization made up of physicians, researchers, public health workers, individuals and other health professionals." The goals of the Network include reducing premature mortality of men, fostering health care education for men and their families, increasing physical and mental health of men, and energizing government involvement in men's health.

The main page of the MHN Web site is quite busy (see Figure 9). There are links to information about MHN and its goals, strategies, and activities, but the heart of the information found here is in the library of documents and links to other Web sites. The "Reference Library" con-

FIGURE 9. Men's Health Network Web Page

tains documents about health topics (e.g., prostate issues, diabetes, cardiovascular diseases) and social science topics (e.g., fathers and parenting, child custody, domestic violence). Adobe Acrobat reader is required for the PDF documents. The "Men's Links" page has an extensive (45 pages when printed out) list of links to Web sites dealing with men's health issues. The links are arranged with general men's health resources listed first, followed by very generalized subjects (e.g., prostate, other cancer sites, domestic violence). While this list of links could be arranged a bit better, it still is useful for consumers to browse through. Additional information accessible from the main page includes Health Scout health news and a photo gallery.

MEDLINEplus: Men's Health Topics
<http://www.nlm.nih.gov/medlineplus/menshealth.htm>

MEDLINEplus is listed as a resource site for almost any health care information topic. Compiled by the U.S. National Library of Medicine, this site links to quality information resources on the Internet, using preexisting information from government agencies and health care organizations. Material is intended for health care consumers.

Much of the information on the "Men's Health Topics" page is cross-referenced to topics such as "Sexually Transmitted Diseases" and "Male Genital Disorders" (see Figure 10). Selecting a topic (e.g., Prostate Diseases), leads to a page of links to information on diagnosis/symptoms, treatment, clinical trials, organizations, statistics, etc. Selecting one of these links then opens up a new window with a direct connection to the information. This inserted window makes it clear that you are connected to another/outside source. Some sources require Adobe Acrobat reader to download the PDF file.

National Women's Health Information Center:
What About Men's Health?
<http://www.4woman.gov/mens/index.cfm>

The last place you might want to look for information about men's health is on a women's health site, right? In reality, women–spouses, friends, daughters–are frequently the ones who look up information about men's health, so what better place to locate this information. "What About Men's Health?" is located in the NWHIC, one of the premier sites for women's health information. Much of the material is supplied by the U.S. government (as indicated by an American flag), and

FIGURE 10. MEDLINEplus: Men's Health Topics Web Page

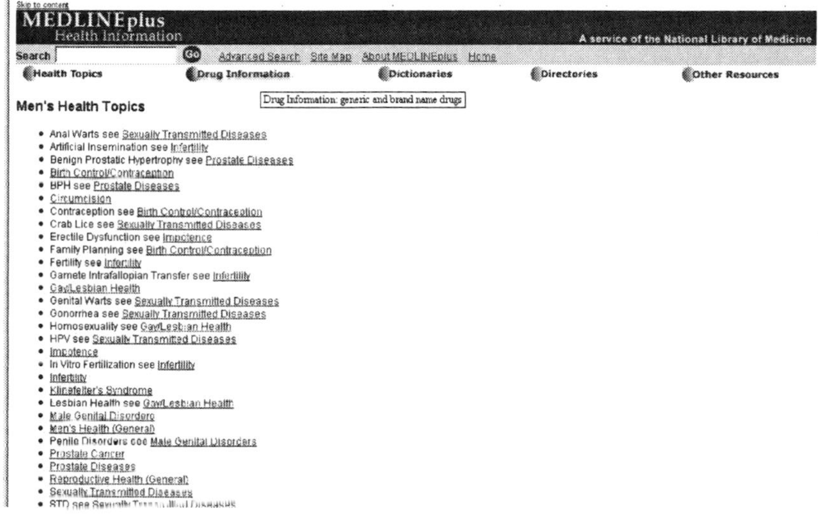

the remainder is provided by reputable sources (e.g., American Cancer Society).

The main page has links to twelve subject categories (e.g., cancer) (see Figure 11). The subject pages include direct links to some publications, plus a further breakdown by subject with more links (e.g., prostate cancer, with links to over 10 documents and 3 organizations). Also from the main men's health page arc links to special groups such as "College Age Men" and "Minority Men."

CONCLUSION

These resources should be used as first-line sites for finding health care information in general. While many more excellent sites are available on the Internet, these sites were singled out as providing quality, non-biased information and as having material specifically geared toward men's health. Other articles in this work will mention many of these sites, plus provide additional sites to find more specific, subject-oriented information.

FIGURE 11. NWHIC: What About Men's Health Web Page

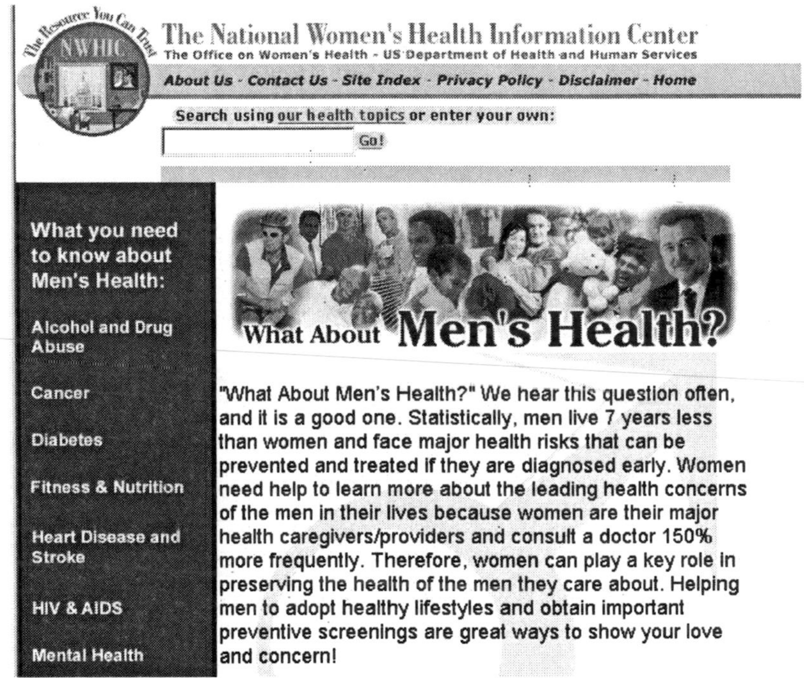

REFERENCES

1. Illinois Department of Aging. "Facts on Aging . . . About Men's Health." Issue 5, May 19, 1999. Available: <http://www.state.il.us/aging/onage05.htm>. Accessed: September 13, 2001.

2. The Male Health Center. "Why Men Don't Go to the Doctor." Available: <http://www.malehealthcenter.com/Bullet.html>. Accessed: September 10, 2001.

3. Illinois Department of Aging. "Facts on Aging . . . About Men's Health." Issue 5, May 15, 1999. Available: <http://state.il.us/aging/onage05.htl>. Accessed: September 13, 2001.

4. U.S. Centers for Disease Control. National Center for Health Statistics. "Fast Stats AtoZ. Men's Health." Available: <http://www.cdc.gov/nchs/fastats/men.htm>. Accessed: September 13, 2001.

5. Fox, Suzannah, and Rainie, Lee. "The Online Health Care Revolution: How the Web Helps Americans Take Better Care of Themselves." Pew Internet and American Life Project. 2000. Available: <http://www.pewinternet.org>. Accessed: September 13, 2001.

6. Reents, S. "Impacts of the Internet on the Doctor-Patient Relationship: The Rise of the Internet Health Consumer." Available: <http://www.cyberdialogue.com/pdfs/wp/wp-cch-1999-doctors.pdf>. Accessed: September 18, 2001.

Information Resources for Men with Cancer

Julie Keany Hodorowski

SUMMARY. Finding accurate information on the Internet can be overwhelming and frustrating, for novice and sophisticated computer users alike. A search for cancer information can turn up thousands of Web sites, not all of which are reliable or well organized. The goal of this article is to point computer users to an array of credible Web sites for cancers that frequently affect men. According to the American Cancer Society, more than 1.2 million Americans will be diagnosed with cancer this year, over 600,000 of them men.[1] United States men have about a 1 in 2 lifetime risk of developing cancer compared with a 1 in 3 risk for women.[1] Cancer is the second leading cause of death in the United States.[1] The Web sites selected address the cancers that occur most frequently and have the highest death rates among men in the United States. *[Article copies available for a fee from The Haworth Document Delivery Service: 1-800-HAWORTH. E-mail address: <getinfo@haworthpressinc.com> Website: <http://www.HaworthPress.com> © 2002 by The Haworth Press, Inc. All rights reserved.]*

KEYWORDS. Cancer Web sites, cancer resources, men and cancer

INTRODUCTION

Many patients diagnosed with cancer want as much information about their disease as they can find. Increasingly, they are turning to the

---

Julie Keany Hodorowski, RN, MA (hodorowj@mskcc.org) is Cancer Information Service Manager, Memorial Sloan-Kettering Cancer Center, 1275 York Avenue, Box 166, New York, NY 10021.

[Haworth co-indexing entry note]: "Information Resources for Men with Cancer." Hodorowski, Julie Keany. Co-published simultaneously in *Health Care on the Internet* (The Haworth Information Press, an imprint of The Haworth Press, Inc.) Vol. 6, No. 1/2, 2002, pp. 17-27; and: *Men's Health on the Internet* (ed: M. Sandra Wood, and Janet M. Coggan) The Haworth Information Press, an imprint of The Haworth Press, Inc., 2002, pp. 17-27. Single or multiple copies of this article are available for a fee from The Haworth Document Delivery Service [1-800-HAWORTH, 9:00 a.m. - 5:00 p.m. (EST). E-mail address: getinfo@haworthpressinc.com].

Internet to find information about cancer diagnosis and treatment, and to locate support groups in their area. But finding accurate information on the Internet can be overwhelming and frustrating, for novice and sophisticated computer users alike. A search for cancer information can turn up thousands of Web sites, not all of which are reliable or well organized. The goal of this article is to point computer users to an array of credible Web sites for cancers that frequently affect men. According to the American Cancer Society, more than 1.2 million Americans will be diagnosed with cancer this year, over 600,000 of them men.[1] U.S. men have about a 1 in 2 lifetime risk of developing cancer compared with a 1 in 3 risk for women.[1] Cancer is the second leading cause of death in the United States.[1] The specific cancers selected for this review are those that occur most frequently and have the highest death rates among U.S. men. While prostate cancer is the leading cancer diagnosed in men today, it will be discussed in a separate article in this issue.

MEN AND CANCER

A recent survey, "Out of Touch: American Men and the Health Care System," financed by The Commonwealth Fund, a private foundation that supports independent research on health and social issues, found that three times as many men as women had not seen a doctor in the previous year and one in three had no regular doctor.[2] One-quarter of the men surveyed reported that they would wait "as long as possible" before seeking medical advice for a health care problem.[3] This reluctance to see a doctor or other medical professional reduces the likelihood that many men will undergo cancer screening or other procedures to detect cancer early, when it is most easily treated. The survey also found that more than half of the males reported that they had not had a physical exam in the previous year; 60% of men age 50 and older had not been screened for colon cancer and 40% had not had a PSA test for prostate cancer in the past year.[3] These screening tests are relatively simple procedures that can detect cancer early.

Dr. Robert Alt commented on the Commonwealth survey by saying that he believes the dire state of men's health is, arguably, the most serious U.S. health problem today.[3] "In the past 35 years, for example, the incidence of female cancer has changed little, while the incidence of cancer in men has increased 20 percent."[3]

Perhaps one way to reach men is through the women in their lives, primarily their wives or daughters. Women are the major health-caregivers/providers for their families, and they consult doctors 150% more frequently than men.[4] Therefore, women can play a key role in preserving the health of the men they care about.[4] Recognizing the influence women have in getting men to seek medical care, the National Women's Health Information Center features a Web page "What About Men's Health?" <http://www.4women.gov/Mens/index.cfm?page=111&text=yes>. This Web page targets men by providing information on specific cancers and tobacco use, and it provides other resources for men's general health.

CANCER INFORMATION ON THE INTERNET

According to the UCLA Internet Report, "Surveying the Digital Future," "the Internet represents the most important technological development of our generation; its effects may surpass those of television and could, over the decades, equal the influence of the printing press."[5] In a report to President Bush, Richard Klausner, M.D., Director of the National Cancer Institute, states "At no other time in history has it been so easy for so many people to access such a vast wealth of information . . . the Internet has multiplied exponentially our ability to make large amounts of information available to a wide audience quickly and easily."[6]

According to Fox and Rainie, when men look for information about a specific illness, they are likely to report that their Web search affected their decisions about how to treat the illness.[7] Men want specific information about what happens to someone during an illness and when certain treatments or drugs are administered.[7]

It is not unusual today for newly diagnosed cancer patients to visit their physician armed with information from cancer-related Web sites. Among those who use the Internet, more than 70% report that the health information they find influences their decision about treatment.[7] The Internet has made it easier and faster for people who live just about anywhere to get access to the best information about cancer.

The number of people seeking health information online continues to grow at twice the rate of overall online users.[8] Approximately one-third of those looking specifically for information about diseases are searching for cancer-related content, while one in four are searching for information on heart disease, making these two illnesses by far the most sought after disease categories.[8] Given that so many patients use the Internet as a decision-making tool, one major challenge is to ensure that people are getting credible and accurate health information. There are growing concerns that much of the information on the Internet may be inaccurate, putting the consumer at unnecessary risk. Although many health professionals agree that the Internet greatly benefits consumers because of the enormous amounts of information available, they are, at the same time, concerned that the poor quality of some information on the Web will undermine informed decision-making.[9, 10] The Web sites listed throughout this article, while not exhaustive, will serve as credible resources for computer users searching for information about cancers that frequently affect men. Information alone can never replace the wisdom of a physician. Patients should discuss with their physicians any information that may affect their decision about treatment. Each cancer case is like a fingerprint, different in every person; therefore, treatment needs to be tailored accordingly.

Professionals in health care and information services struggle to keep up as cancer-related Web sites grow exponentially in number. The volume of information on the Internet is so vast that individual search engines cover only 3% to 34% of all indexible Web sites.[11] Sometimes, the reason why search engines have retrieved a document is not immediately apparent to the searcher. The Internet has been likened to a huge public library with no card catalog, which can be very intimidating to searchers.

EVALUATION OF MEDICAL WEB SITES

Throughout the following pages an attempt has been made to provide the most accurate up-to-date Web sites for men with cancer. The intent is to ease the burden computer users face when searching the Internet for cancer information by pointing them to specific Web sites. Many of the Web sites have information available in Spanish, but unfortunately, resource materials are limited. The sites selected for this article were based on literature reviews and personal experience working in the field of cancer communication. Several organizations have developed review criteria helpful in the evaluation of Web site content. Two of these organizations, Health On the Net Foundation (HON) <http://www.hon.ch/ HONcode/Conduct.html> and the Science Panel on Interactive Communication and Health <http://www.health.gov/scipich/IHC/checklist.htm>, were used in selecting Web sites and resource links for this article. Government Web sites and Web sites that meet the criteria of the HON and/or SciPICH are cited in this article. The following resources are provided for informational purposes only; the author does not assume responsibility for their content.

WEB SITES FOR MEN CONCERNED ABOUT CANCER

adam Health Illustrated Encyclopedia
<http://www.nlm.nih.gov/medlineplus/encyclopedia.html>

This site is useful for men who want to see illustrations as well as text to better understand diagnostic tests such as colonoscopy, anatomy, and some surgical procedures. It contains an extensive library of medical photographs and illustrations.

American Cancer Society
<http://www.cancer.org>

The ACS Web site was recently redesigned and has a wealth of information for men on a broad range of topics including an interactive treatment decision tool and a bookstore. Questions can be answered via an e-mail form on the Web site.

American College of Physicians: Home Care Guide for Advanced Cancer
<http://www.acponline.org/public/h_care/contents.htm>

This *Home Care Guide for Advanced Cancer* has been designed for use by patients, family caregivers, hospice workers, and other health providers. This practical guide is for those preparing to confront the dying process. The problem solving approach is helpful in providing a step-by-step approach to common issues faced in coping with advanced cancer.

American Society of Clinical Oncology
<http://www.asco.org/people/html/m_people.htm>

The information on ASCO OnLine represents the very latest in the diagnosis and treatment of cancer. Oncologists have developed the Web site. The page "Find an Oncologist" <http://www.asco.org/people/db/html/m_db.htm> enables men to search the ASCO member database for the names and affiliations of more than 15,000 oncologists worldwide. "The People Living with Cancer" page <http://www.asco.org/people/html/m_people.htm> is written in easy-to-understand language and is devoted to helping men and their loved ones better understand cancer and its causes, as well as treatments. Since colon and lung cancers are common in men, the "Patient's Guide: Follow-Up Care for Colorectal Cancer" <http://www.asco.org/people/rs/html/m_patguide color.htm> will be helpful to those seeking guidelines for follow-up care. The "Patient's Guide: Advanced Lung Cancer Treatment" <http://www.asco.org/people/rs/html/m_patguidelung.htm> will provide guidelines to men with inoperable non-small-cell lung cancer and describes various treatments.

Association of Cancer Online Resources, Inc.
<http://www.acor.org>

Men interested in communicating with others with cancer will find electronic mailing lists and unique Web sites for Internet relay chat on this site.

Cancer Research Foundation of America–Men's Health
<http://preventcancer.org/mhealth.cfm>

The Cancer Research Foundation of America men's health page provides age specific screening and prevention information for colorectal, lung, skin, prostate, and testicular cancers.

Centers for Disease Control and Prevention–Men's Menu
<http://www.cdc.gov/health/mensmenu.htm>

The CDC provides resources on tobacco use and colon, skin, and prostate cancer. "The Colorectal Cancer Prevention and Control Initiative" <http://www.cdc.gov/cancer/colorctl/colorect.htm> page contains resources on the importance of prevention and early detection. "The National Skin Cancer Prevention Education Program" <http://www.cdc.gov/cancer/nscpep/skin.htm> page contains resources on the highly curable basal cell and squamous cell carcinomas and the more serious malignant melanoma.

ClinicalTrials.gov
<http://clinicaltrials.gov>

Developed by the National Institutes of Health (NIH) through its National Library of Medicine, this site mirrors the content of the cancerTrials site, but may have other federal agency trials included. This site is for men interested in exploring clinical research studies.

FDA Oncology Tools
<http://www.fda.gov/cder/cancer/index.htm>

Oncology Tools contains a variety of information related to cancer and approved cancer drug therapies. It is particularly helpful if men are looking for specific information on newly approved cancer drugs and information on the use of cancer drugs for special exception or compassionate use.

healthfinder®
<http://www.healthfinder.gov/justforyou/>

healthfinder® is a free guide to reliable consumer health and human services information, developed by the U.S. Department of Health and Human Services. Men can go to the "Just for you" section specific to men and select from an alphabetical list to explore the cancer topic resources.

Johns Hopkins Medical Institutions Pancreas Cancer Home Page
<http://pathology.jhu.edu/pancreas>

The Department of Pathology at the Johns Hopkins Hospital maintains the unique Pancreatic Cancer Web site for men interested in treatment options, research, and literature on pancreatic cancer. The site allows men the option of using an active chat room where users can post messages and read and reply to all previously posted messages.

Leukemia and Lymphoma Society
<http://www.leukemia-lymphoma.org>

This site provides information on leukemia, lymphoma, and multiple myeloma. Personalization is a unique feature of this Web site.

Lymphoma Research Foundation of America
<http://www.lymphoma.org>

A buddy program is available to match newly diagnosed patients with other lymphoma patients who have coped with the disease.

MEDLINEplus
<http://medlineplus.gov>

This site is a service of the National Library of Medicine that provides up-to-date, quality health care information from the world's largest medical library, the National Library of Medicine at the NIH. MEDLINE does not provide the full text of journal articles. One of the more valuable sections for men is the U.S. Pharmacopeia's Drug Information, Vol. II, and "Advice for the Patient."

Memorial Sloan-Kettering Cancer Center–Comprehensive Patient Education
<http://wwwpe.mskcc.org/PatientEd/Course.nsf>

This is a multi-media program containing overviews on specific cancers including those commonly occurring in men. The overviews include information about diagnosis and treatment. It is best accessed from a computer with high-speed Internet access.

National Cancer Institute
<http://www.cancer.gov/>

Since the National Cancer Institute (NCI) has several Web sites and resources available online, they will be discussed separately. The vast majority of cancer-related Web sites retrieved via search engines provide information from the NCI. Many of these secondary sites do not update their links to NCI material. It is important to go directly to the primary NCI resources to ensure accuracy. Recently the NCI added LiveHelp to assist users in navigation and to respond to Instant Messaging via a live chat. This service is available Monday-Friday, 9 a.m.-5 p.m. ET.

Cancer Information Service
<http://cis.nci.nih.gov/>

This NCI Web site has a cancer resources page that includes the "Cancer Facts" collection of fact sheets that address a variety of cancer topics. The content and contact information is highly reliable. Some of the topics of possible interest to men include *Cancer Information Sources* <http://cis.nci.nih.gov/fact/2_1.htm>; *National Organizations That Offer Services to People with Cancer and Their Families* <http://cis.nci.nih.gov/fact/8_1.htm>; and *The National Cancer Institute Cancer Centers Program* <http://cis.nci.nih.gov/fact/1_2.htm>. Men with financial concerns may be interested in *Questions and Answers: Cancer Studies at the Warren Grant Magnuson Clinical Center* <http://cis.nci.nih.gov/fact/1_22.htm>, the federal research hospital in

Bethesda that provides cancer treatment in clinical trials at no cost to patients.

There is a link to the NCI's Publications Locator for men who wish to locate and order free NCI publications online <https://cissecure.nci.nih.gov/ncipubs/>. An example of an online publication is *Pain Control: A Guide For People with Cancer and Their Families* available in English, Spanish, and large print <http://oesi.nci.nih.gov/paincontrol/index.html>.

CancerNet™
<http://cancernet.nci.nih.gov>

The NCI's PDQ® and CANCERLIT® databases are included in CancerNet™. The NCI's PDQ® database contains the latest information on over 100 different types of cancer and a link to a Cancer Dictionary <http://cancernet.nci.nih.gov/dictionary.html>. Men can find treatment, screening, prevention, genetics, and supportive care information here. Some of the specific pages for common cancers in men include: colon and rectal cancer <http://cancernet.nci.nih.gov/Cancer_Types/Colon_And_Rectal_Cancer.shtml>; lung cancer <http://cancernet.nci.nih.gov/cancer_types/lung_cancer.shtml>; skin cancer <http://cancernet.nci.nih.gov/cancer_Types/Skin_Cancer.shtml>; non-Hodgkin's lymphoma <http://cancernet.nci.nih.gov/cancer_types/lymphoma. shtml>; leukemia <http://cancernet.nci.nih.gov/Cancer_Types/Leukemia.shtml>; and pancreatic cancer <http://cancernet.nci.nih.gov/Cancer_Types/Pancreatic_Cancer.shtml>.

cancerTrials
<http://cancertrials.nci.nih.gov>

Access to information regarding clinical trials is of critical importance to all cancer patients. The premier Web site is cancerTrials, NCI's comprehensive clinical trials site, providing access to NCI's clinicalTrials database, news about cancer research, and resources for patients and health care professionals about participating in clinical trials.

University of Michigan Health Systems Patient Education Resource Center Pathways
<http://www.cancer.med.umich.edu/learn/percpathways.htm>

The Patient Education Resource Center (PERC) of the Cancer Center at the University of Michigan developed PERC Pathways to help newly diagnosed cancer patients and their loved ones find sources of information and support. Resources included are links to Web sites and suggestions for books, a newsletter, and videos. The resources are well organized, but some are specific to Michigan. Men with colon or rectal cancer should go to <http://www.cancer.med.umich.edu/learn/pwcolon.htm#>; and for lung cancer to <http://www.cancer.med.umich.edu/learn/pwlung.htm>.

CONCLUSION

The demand for cancer information will continue to increase as the population ages. Men with cancer face increasingly complex medical decisions, and they will look to the Internet to assist them in locating information. Hopefully, the resources provided in this article will help empower men to understand their disease and make informed choices about their care.

REFERENCES

1. American Cancer Society. *Cancer Facts and Figures 2001*. Available: <http://www3.cancer.org/eprise/main/docroot/STT/content/STT_1x_2001_Facts_and_Figures.pdf>. Accessed: August 20, 2001.

2. Sandman, David; Simantov, Elisabeth; and An, Christina. The Commonwealth Fund. "Out of Touch: American Men and the Health Care System." (March 2000). Available: <http://www.cmwf.org/publist/publist2.asp?CategoryID=10>. Accessed: August 17, 2001.

3. Alt, Robert. "Men's Health Matters: A Gender in Crisis." *Wisconsin Medical Journal* 100 (July 16, 2001). Available: <http://www.wismed.org/wmj/v100n03/menshealth.htm>. Accessed: August 17, 2001.

4. National Women's Health Information Center (NWHIC) Available: <http://www.4woman.gov/mens/index.cfm>. Accessed: August 17, 2001.

5. Lebo, H. "The UCLA Internet Report: Surveying the Digital Future." (November 2000). Available: <http://www.ccp.ucla.edu/ucla-internet.pdf>. Accessed: August 17, 2001.

6. National Cancer Institute. "Scientific Priorities for Cancer Research: NCI's Extraordinary Opportunities." (February 12, 2001). Available: <http://plan.cancer.gov/scpcomms.htm>. Accessed: August 17, 2001.

7. Fox S., and Rainie, L. "The Online Health Care Revolution: How the Web Helps Americans Take Better Care of Themselves." Washington, DC: Pew Charitable Trusts, November 26, 2000. Available: <http://www.pewinternet.org/reports/toc.asp?Report=26>. Accessed: August 17, 2001.

8. Reents, S. "Impacts of the Internet on the Doctor-Patient Relationship: The Rise of the Internet Health Consumer." Available: <http://www.cyberdialogue.com/pdfs/wp/wp-cch-1999-doctors.pdf>. Accessed: August 17, 2001.

9. Thomas, R., Eng, D. and Gustafson, H. "Science Panel on Interactive Communication and Health. Wired for Health and Well-Being: the Emergence of Interactive Health Communication." Washington, DC: U.S. Department of Health and Human Services, US Government Printing Office, April 1999. Available: <http://www.health.gov/scipich/pubs/finalreport.htm>. Accessed: August 17, 2001.

10. Department of Health and Human Services. *Healthy People 2010* Available: <http://www.health.gov/healthypeople/document/html/volume1/11healthcom.htm>. Accessed: August 17, 2001.

11. Lawrence S., and Giles, C. "Searching the World Wide Web." *Science* 280(April 3, 1998): 98-100.

Prostate Cancer, Prostatic Hypertrophy, and Prostatitis: A Select List of Resources on the Web

A. James Bothmer

SUMMARY. The prostate is a male sex gland that helps produce semen. There are several serious conditions associated with the prostate. Benign prostatic hypertrophy (BPH), prostate cancer, and prostatitis will be briefly defined. Education and self-help sources that raise awareness, which in turn can lead to prevention and successful treatment, are discussed. The World Wide Web contains huge amounts of information. The following article includes a select list of quality sites dealing with some aspect of the prostate. The author intends for readers to use this as a starting point for information relevant to those suffering from prostate disease or cancer or who just want to know about the subject. *[Article copies available for a fee from The Haworth Document Delivery Service: 1-800-HAWORTH. E-mail address: <getinfo@haworthpressinc.com> Website: <http://www.HaworthPress.com> © 2002 by The Haworth Press, Inc. All rights reserved.]*

KEYWORDS. Benign prostatic hypertrophy, prostate cancer, prostatitis, prostatic hyperplasia, Internet, health resources for men

---

A. James Bothmer, MALS, AHIP (jbothmer@creighton.edu) is Director, Creighton University Health Sciences Library, 2500 California Plaza, Omaha, NE 68178.

[Haworth co-indexing entry note]: "Prostate Cancer, Prostatic Hypertrophy, and Prostatitis: A Select List of Resources on the Web." Bothmer, A. James. Co-published simultaneously in *Health Care on the Internet* (The Haworth Information Press, an imprint of The Haworth Press, Inc.) Vol. 6, No. 1/2, 2002, pp. 29-37; and: *Men's Health on the Internet* (ed: M. Sandra Wood, and Janet M. Coggan) The Haworth Information Press, an imprint of The Haworth Press, Inc., 2002, pp. 29-37. Single or multiple copies of this article are available for a fee from The Haworth Document Delivery Service [1-800-HAWORTH, 9:00 a.m. - 5:00 p.m. (EST). E-mail address: getinfo@haworthpressinc.com].

INTRODUCTION

Following are brief descriptions of prostate cancer, benign prostatic hypertrophy (BPH), and prostatitis. In all cases, the reader should consult his doctor if there are any questions about symptoms or these conditions.

The prostate is a male sex gland that helps make semen. It sits below the bladder and wraps around the urethra like a donut. The urethra is the tube that carries urine from the bladder into the penis. The urine is then passed out of the body. As the prostate grows, it can squeeze the urethra, making it harder to pass urine. This can block urine flow and cause urinary problems.[1] Benign prostatic hypertrophy or BPH is very common in men over 45 years of age, but rarely causes problems before age 60. It is not known what causes BPH. It may be just part of getting older, or it may be caused by a change in hormones as men age.[1]

Prostate cancer starts as a tumor or lump. Normal cells split in a planned way, making more cells only when needed. Cancer happens when cells grow and divide without control or order, often making too much tissue (tumor). Cancer cells may grow into nearby healthy tissue. They may also break away from the tumor. These cells travel through the blood stream or lymphatic system to other parts of the body. This is called metastasis when the tumor cells spread to other areas of the body. While specific causes of prostate cancer are still not clear, it is known that you cannot catch it from someone else. There are, however, factors that increase the risk of developing this cancer. They are:

- Eating a steady diet of foods with high fatty content
- Engaging in sexual contact with many people or getting venereal diseases
- Working around chemical cadmium or other known cancer causing chemicals
- Being a man over the age of 50 means having a higher risk
- Being an African American man, who have the highest rate of prostate cancer in the United States
- Having a family history of prostate cancer increases risk.[2]

Prostatitis is an inflammation of the prostate gland. It can cause intense pain, urinary complications, sexual dysfunction, infertility, and a reduced quality of life. Not a great amount of research exists on the topic. There is, however, no evidence that prostatitis leads to prostatic cancer and BPH. An entire foundation has been established to sponsor and encourage research and to provide reliable information on this condition.[3]

METHODOLOGY

Resources on the Web are increasing at an exponential rate. There is simply an overwhelming amount of information. For example, a simple search on the Google search engine <http://www.google.com> using the words "prostate cancer" yielded over 800,000 hits! Obviously there were many duplicate links and much overlap, but how would one know which sources to consult first for accurate information? In addition, the reading level of many of these sites is very high. In a recent article in the *Journal of the American Medical Association,* the authors found that less than one quarter of a search engine's first pages of links led to relevant content. On an average, less than 50% of the clinical elements on Web sites were more than minimally covered and completely accurate, and all English sites required a high school level or greater reading ability.[4]

Essentially, there are four major sources of health related information repositories on the Web: governmental agencies (federal, state, local); academic-related organizations; professional organizations; and, increasingly, advocacy organizations. Most of the sites mentioned in this paper are .org, .edu, or .gov sites. There are many .com sites and some are listed, but the consistency of quality tends to be greater in the other three sites.

When evaluating a Web site, the best criteria, in this author's opinion, are in an article done by Kim et al.[5] The twelve criteria listed in this article for determining a quality site are:

- Content of site
- Design and esthetics
- Disclosure of authors, sponsors, developers
- Currency of information
- Authority of source
- Ease of use
- Accessibility and availability
- Links
- Attribution and documentation
- Intended audience
- Contact addresses or feedback mechanism
- User support.

Following is a highly selective list of sites that this author considers to be good resources for obtaining information about where to go for help if suffering from prostate related diseases, or if information is required. They are not presented in order or any level of relevancy. The U.S. federal government is almost always an excellent source of health information on the Web, which makes it a good place to begin.

There are several excellent governmental sponsored sites through the National Institutes of Health (NIH). These sites are a must stop, especially if no other Web site is accessed. Your tax dollars are truly at work here.

WEB SITES ON THE PROSTATE

CancerNet: Prostate Cancer <http://cancernet.nci.nih.gov/wyntk_pubs/prostate.htm>

This is an excellent source of information about prostate cancer and other diseases of the prostate. It is actually an online booklet about cancer of the prostate and is laid out in a logical format with easy links to specific chapters. A dictionary is included to help the reader understand terminology, and some of the more difficult words in the text are directly linked to the dictionary. There are also several good illustrations that help the reader understand the function of the prostate gland. Although the reading level may be high, this is one of the best sources for information available on prostate diseases.

MEDLINEplus and PubMed <http://medlineplus.nlm.nih.gov/> and <http://www.ncbi.nlm.nih.gov/entrez/query.fcgi>

Of course, no search would be complete without visiting MEDLINEplus and PubMed, both available through the National Library of Medicine. Under Health Topics in MEDLINEplus <http://www.nlm.nih.gov/medlineplus/healthtopics_p.html> there are excellent links to prostate cancer at <http://www.nlm.nih.gov/medlineplus/prostatecancer.html> and prostate diseases at <http://www.nlm.nih.gov/medlineplus/ prostatediseases.html>. Upon clicking on either of these links, the user is presented with a vast array of additional linked information. Included are overviews, clinical trials, diagnosis and symptoms, treatments, directories, dictionaries, and statistics. Many of the documents are available in Spanish as well as English. No search would be complete without mentioning the National Library of Medicine's research bibliographic database, PubMed. PubMed searches over 4,000 biomedical journals for information on many medically related subjects. This database can certainly be used by the layperson, but is geared toward the health care professional.

Clinical Trials
<http://www.clinicaltrials.gov/>

Clinical Trials is another good point of reference on the NIH site. This site provides information on current research. A simple search on prostate yields 122 documents. Each document indicates whether the researcher is recruiting or not yet recruiting patients for the study.

Dirline
<http://dirline.nlm.nih.gov/>

This is an additional source of information on organizations that deal with the prostate.

LocatorPlus
<http://locatorplus.gov/>

LocatorPlus is the government's online public access catalog of books, journals, and audiovisuals. A search on "prostate" yields numerous "hits." Most of the items reside at the National Library of Medicine and can be borrowed through a public library. This useful database also allows the user to search other NIH databases.

National Institute of Diabetes, Digestive, and Kidney Disease
<http://www.niddk.nih.gov/>

This site includes 88 documents on the prostate ranked by relevancy and ranging from news releases to research documents. It is not the best of the NIH sites, but still deserves a look.

National Human Genome Research Institute
<http://www.nhgri.nih.gov/>

This site contains 94 documents dealing with the prostate, with items relevancy ranked. It is more research oriented, especially towards genetics research.

Administration on Aging
<http://www.aoa.dhhs.gov/>

This government site contains some excellent information on prostate health in the aging population.

Prostatitis Foundation
<http://www.prostatitis.org>

This site is sponsored by an unrestricted grant from Merck. Information is presented about sources in Spanish, English, Dutch, Swedish, and Canadian. The Prostatitis Foundation, founded in 1995 and located in Smithshire, Illinois, was set up to encourage research, provide information, and collect data. They have sponsored four national conferences to date. Pages within this site include a moderated e-mail distribution list <http://www.prostatitis.org/moderated.html>, and a chat room <http://www.prostatitis.org/chatroom.html>. The site exists for patients and is maintained by volunteers. It is broken into three areas: Prostatitis, Prostate Cancer, and Benign Prostatic Hypertrophy (BPH). Each area has its own Web site. The Prostatitis site is part of the Prostatitis Foundation located in Illinois <http://www.prostatitis.org/aboutpf.html> and includes research information. The BPH site is interactive and allows the individual to enter personal information in an online survey and discuss conditions in chat room format with people from around the world. Prostate.org leads to three separate Web sites on BPH, Prostate Cancer, and Prostatitis. Prostate Cancer site links to the USToo site (see description elsewhere in this article).

American Cancer Society Prostate Cancer
<http://www3.cancer.org/cancerinfo/load_cont.asp?ct=1>

Another excellent source for information on prostate cancer is the American Cancer Society's Prostate Cancer Resource Center. In addition to containing good information from a very reliable source, there is a Prostate Cancer Profiler page, which allows the user to determine how much at risk he is.

Prostate Health Resources
<http://www.prostate90.com/>

Alternative and complementary medicine are terms increasingly used to describe non-traditional medical approaches to medical problems. While this author is not advocating alternative medical approaches to therapy, this site does contain good information. It would be my recommendation that readers who want to take this approach consult with their physicians before doing so. The site does have a discussion group and is monitored by Larry Clapp, author of *Prostate Health in 90 Days*.

National Prostate Cancer Coalition
<http://www.4npcc.org/>

The National Prostate Cancer Coalition is an awareness, outreach, and advocacy group. This site contains a good source of information about how to lobby for increased funding for research on prostate cancer and related prostate diseases.

Us Too International
<http://www.ustoo.com/>

This site claims to be the world's largest independent, charitable network of support groups concerning prostate issues. One of their goals is to foster public awareness of prostate diseases. There is a daily news feature that readers can register to receive.

Web MD
<http://my.webmd.com/>

This is a good, reliable source of health information with some excellent resources on the prostate included in the site.

Cancer Links
<http://www.cancerlinks.org/>

Cancer Links may be one of the most useful tools for locating information on all aspects of the prostate. It is a Web site that has links to hundreds of screened sites on the prostate. One of the unique features on this site is that the reader can select from sixteen languages and is then directed to Web sites in those languages/countries. While many of the sites are in English, there is often a translation button, which converts the site to the language of origin. There are over 40 sites devoted to the prostate.

Urology Network
<http://www.urologynet.org>

The Urology Network is a global professional meeting place and information source on all aspects of urology. It is targeted to the professional, but there is extensive information on the prostate that is worth reviewing. This network is sponsored by the Societe Internationale d'Urologie.

The American Urologic Association <http://www.auanet.org>

This professional association dedicated to the "advancement of urologic care" contains an excellent brochure titled "Prostate Cancer Guide for Men, A Doctor's Guide for Patients" <http://www.auanet.org/patient_info/index.cfm>.

American Prostate Society <http://www.ameripros.org/>

This society, located in Maryland, is a non-profit society devoted to reducing the suffering and deaths caused by prostate diseases. It appears not to be recently updated and some pages are "under construction." It might be worthwhile to bookmark it and check it out periodically for progress.

The Prostate Help Association <http://www.pha.u-net.com>

The Prostate Help Association is a source of information on prostate cancer, prostatitis, and benign prostatic hyperplasia. The site includes information on tests, treatments, and an overview of prostate conditions. A newsletter is available for a subscription.

Association for the Cure of Cancer of the Prostate or CaP CURE <http://www.capcure.org>

This site contains excellent publications in PDF format as well as good information on clinical trials. There is also a list of charity events in support of prostate cancer.

EDUCATIONAL SITES

Educational institutions (those Web sites ending with .edu) contain a wealth of information on health related topics. Several that yielded excellent results on prostate health include the University of Iowa's Virtual Hospital <http://www.vh.org/>, the Mayo Clinic's MayoClinic.com <http://www.mayohealth.org/home>, and the Johns Hopkins University's Infonet <http://infonet.welch.jhu.edu>. Powered by Ultraseek

Server search software, this last site contains many excellent documents on prostate health and diseases. The University of Pennsylvania's OncoLink <http://cancer.med.upenn.edu/> is another good ".edu" site. For pure research on prostate related health topics, the Sloan Kettering Cancer Institute is an excellent site <http://www.mskcc.org/>. There are many more valuable sites among the nation's more than 120 academic health sciences centers. The above sites are merely representative of the many links available.

CONCLUSION

Other than skin cancer, prostate cancer is the most commonly diagnosed cancer in men. It is estimated that of all men diagnosed each year with cancer, prostate cancer is the diagnosis in 25% of them. It is also a fact that with early diagnosis, prostate diseases including cancer are increasingly treatable. With education, many prostate related problems are also preventable. It is hoped that the links mentioned in this article to prostate information sources on the Internet will help in providing education and awareness for everyone seeking to understand the prostate.

REFERENCES

1. The CareNotes™ System, MICROMEDEX. "The CareNotes™ System: Benign Prostatic Hypertrophy." Englewood, CO (v. 19, expires June 2001). Accessed: May 16, 2001.

2. The CareNotes™ System, MICROMEDEX. "The CareNotes™ System: Prostate Cancer." Englewood, CO. (v. 19, expires June 2001). Accessed: May 16, 2001.

3. Prostatis (2000). "About the Prostatitis Foundation." Available: <http://www.prostatitis.org/>. Accessed: May 16, 2001.

4. Berland, G.; Elliott, M. et al. "Health Information on the Internet." *JAMA* 285:2612-21, 2001. Available: <http://jama.ama-assn.org/issues/v285n20/rfull/joc02274.html>. Accessed: May 31, 2001.

5. Kim, Paul; Eng, Thomas R. et al. "Published Criteria for Evaluating Health Related Web Sites: Review." *British Medical Journal* 318(March 6, 1999):647-9.

Male Infertility and Impotence Resources on the Web

Rebecca A. Danner
Nita K. Splittorff

SUMMARY. This article focuses on infertility and impotence Internet resources for men. As these subjects can be sensitive in nature, the Internet can be an invaluable resource to help facilitate the research process. The selected Internet resources in this article will guide users to some of the best information available at this time. *[Article copies available for a fee from The Haworth Document Delivery Service: 1-800-HAWORTH. E-mail address: <getinfo@haworthpressinc.com> Website: <http://www.HaworthPress.com> © 2002 by The Haworth Press, Inc. All rights reserved.]*

KEYWORDS. Male infertility, impotence, Internet

INTRODUCTION

Recent years have seen an increase in publicity about men's sexual health. Mainstream advertising, new treatments, and a savvier health care consumer have led to increased information on the Web. These de-

Rebecca A. Danner, MSLIS, is Medical Librarian, Health Sciences Library, Maricopa Integrated Health System, 2601 East Roosevelt Street, Phoenix, AZ 85008. Nita K. Splittorff, MSLIS, is Medical Librarian, Health International, 14770 North 78th Way, Scottsdale, AZ 85260.

[Haworth co-indexing entry note]: "Male Infertility and Impotence Resources on the Web." Danner, Rebecca A., and Nita K. Splittorff. Co-published simultaneously in *Health Care on the Internet* (The Haworth Information Press, an imprint of The Haworth Press, Inc.) Vol. 6, No. 1/2, 2002, pp. 39-45; and: *Men's Health on the Internet* (ed: M. Sandra Wood, and Janet M. Coggan) The Haworth Information Press, an imprint of The Haworth Press, Inc., 2002, pp. 39-45. Single or multiple copies of this article are available for a fee from The Haworth Document Delivery Service [1-800-HAWORTH, 9:00 a.m. - 5:00 p.m. (EST). E-mail address: getinfo@haworthpressinc.com].

velopments have prompted a more open dialogue between men and their health care providers, as well as increased men's need to seek more information to improve their quality of life.[1,2,3] In addition, the advances and realization that male infertility can now be treated has given new hope to couples trying to conceive. With new treatments under investigation, more consumer health information on both of these topics is foreseeable.

In searching for quality infertility and impotence sites on the Internet, it was discovered that the majority of men's health content is a subset of larger sites. Much of the Web-based information on male infertility was presented alongside the topic of female infertility. Since fertility can be considered a "couple" issue, men might also find the general infertility information useful. While only one Web site dedicated entirely to impotence is listed, most of the other Web links are to specific pamphlets and articles that were part of larger sites. The quality of content of these specifically linked articles merited their inclusion. Although there are many Web sites dedicated to the topics of infertility and impotence, the resources below are selective in scope.

INFERTILITY RESOURCES

Infertility is commonly thought of as only a female problem. But in fact, 35% of infertility cases are actually due to male problems, while 35% are because of female problems. Meanwhile, 20% are caused by combined problems, and 10% are from unknown causes.[4] Although male infertility is the focus of this article, many of these Web sites concern infertility in general and can inform men of the issues they may face as a couple trying to conceive. These issues include such topics as adoption, coping, and treatments for both themselves and their partners.

American Society for Reproductive Medicine (ASRM) <http://www.asrm.org/Patients/mainpati.html>

The American Society for Reproductive Medicine (ASRM) has a section of its Web site specifically dedicated to resources for patients. In addition, there are links to locate ASRM physicians and those physicians specializing in infertility issues. Links to fact sheets cover many specific topics including Intracytoplasmic Sperm Injection, Sexual Dysfunction, Stress and Infertility, Testing for Male Infertility, Counseling and Support, and more. Although some entries have not been updated in the last five years, there are still a wide variety of topics covered in the fact sheets, patient information booklets, and Frequently Asked Questions (FAQs). This Web site would be much improved if the medical content was reviewed more frequently.

InterNational Council on Infertility Information Dissemination–Infertility Information and Support
<http://www.inciid.org/>

The INCIID is a non-profit organization that aims to provide consumer health information on a variety of infertility topics. This Web site includes interactive Support Forums (chat rooms) where consumers can share information, and Medical Forums that are mediated by health care professionals. Topics include a Men's Room in the Support Forum and a Male Infertility section in the Medical Forum. There is also a Medical Forum dedicated to Spanish speaking users. In addition, there are many fact sheets and reading lists for users to browse, including basic infertility information, an infertility glossary, a semen analysis chart, and more. There are many sections with content authored by physicians, as well as sections excerpted from book chapters. Although there is a link to professional directories, the health care professional listings were not extensive and many missing links occurred during navigation. The initial link to the Reproductive News looked promising, but subsequent links to various news stories were outdated and no longer active. The news stories listed were all at least three months old when this article was written. Despite these negatives, this is quite an extensive site with much valuable information.

MEDLINEplus–Infertility
<http://www.nlm.nih.gov/medlineplus/infertility.html>

MEDLINEplus is a great resource to begin researching any consumer health issue. The fertility subcategory has an excellent listing of male and female aspects of fertility, which include links to overviews, clinical trials, research, treatment, organizations, and more. Some Web page links specifically for men include a Patient's Guide to Low Testosterone, Diagnostic Testing for Male Factor Infertility, and Male Infertility: Causes and Remedies. Also included on the main site is the adam.com *Health Illustrated Encyclopedia* with an entry on infertility that contains anatomical illustrations of the male and female reproductive systems. This Web site also has excellent lay language entries on fertility drugs provided by thc *United States Pharmacopeia (USP) DI Advice for the Patient*. Most MEDLINEplus health topics now have sections with links in Spanish; unfortunately only one is currently included for infertility.

RESOLVE: The National Infertility Association
<http://www.resolve.org>

RESOLVE, a non-profit organization, aims to connect people with infertility resources nationwide. This Web site includes basic infertility information including fact sheets and articles on topics such as where to start, how to select an infertility physician, and myths and facts. A bulletin board is available for interested people to share their stories; however when this article was written, this was limited in the availability of archived posts. Links to clinical trials and local chapters of RESOLVE are available. In addition, a user can join the organization for an annual fee, which provides access to additional services beyond their Web site, including physician referrals and support networks.

Urology Channel–Male Infertility
<http://www.urologychannel.com/maleinfertility/index.shtml>

The Urology Channel is a medical specialty site owned by healthcommunities.com with a mission to provide "physician developed and monitored health communities." Those physicians who are involved in the review board process are listed on the site. Within the physician-reviewed content is a section devoted to male infertility that includes sections on diagnosis, lab tests, sperm tests, causes, risk factors, and more. In addition to the health content, scheduled physician chats are available with archived transcripts of past talks. This site offers a service called MDLocator to help locate urologists in one's geographic area. However, when this article was written, only a limited number of urologists were retrieved.

IMPOTENCE

Male impotence, or erectile dysfunction, is a distinct and widespread problem. Although it can affect men of any age, 52% of men aged 40 to 70 have experienced some degree of erectile dysfunction.[5] Since the likelihood of men encountering sexual dysfunction is great, there is a definite need for consumer health information on this topic. The Internet has provided men an avenue to seek this. In addition, increased publicity with new treatments such as Viagra (Sildenafil) has increased public awareness and, as a result, more information is available to men.

American Medical Association (AMA)–Erectile Dysfunction <http://www.medem.com/MedLB/article_detaillb.cfm?article_ID=ZZZKV6W3NAC&sub_cat=57>

Medem provides the materials from many of the nation's leading medical societies, including the American Medical Association (AMA). The AMA has written an extensive patient information page available on the Medem site regarding erectile dysfunction (ED). It includes information on what ED is, what causes it, how it is diagnosed, and available treatment options.

Mayo Clinic–Erectile Dysfunction <http://www.mayoclinic.com/home?id=DS00162>

The Mayo Clinic provides a superb entry on erectile dysfunction and its treatment. Within the content, there are many links to other related topics on the Mayo Clinic Web site, including information on the topics of physical activity and impotence. A link to the Men's Health section also provides access to other information on topics such as Viagra.

MEDLINEplus–Impotence <http://www.nlm.nih.gov/medlineplus/impotence.html>

The impotence subcategory of MEDLINEplus has the same great features as the Infertility section of MEDLINEplus described above. Included in the main site is adam.com's *Health Illustrated Encyclopedia* with entries on Impotence, Impotence and Age, and Drugs That May Cause Impotence. Overall, this is an excellent starting place for men to locate information.

National Institute of Diabetes and Digestive and Kidney Diseases (NIDDK)–Impotence <http://www.niddk.nih.gov/health/urolog/pubs/impotnce/impotnce.htm>

National Institute of Diabetes and Digestive and Kidney Diseases is a branch of the National Institutes of Health. The online pamphlet on impotence offers a great source for general information on this topic. In addition, the graphics and referrals to other resources are quite helpful. The Spanish version of this pamphlet is also available at <http://www.niddk.nih.gov/health/urolog/pubs/imptcia/index.htm>.

Sexual Function Health Council of the American Foundation for Urologic Disease, Inc.
<http://www.impotence.org>

The navigation on this Web site is very straightforward. The entire site is dedicated to information on impotence and is reflected heavily in the content. The Ask the Expert allows the user to e-mail questions to a male nurse. There is also a chat room for users to communicate with others regarding this health issue. In addition to these interactive features, the health content includes sections on Confronting Impotence, Diagnosis, FAQs, Information for Partners, and allows visitors to take an ED Quiz.

Urology Channel–Male Infertility
<http://www.urologychannel.com/erectiledysfunction/index.shtml>

The Urology Channel, mentioned above in the Impotence listing of Internet resources, also has information on sexual dysfunction. The site offers the same type of information and features previously described.

CONCLUSION

With the ever-increasing amount of consumer health information available on the Internet, there has also been a new focus and awareness on men's health issues. These selected sites on male infertility and impotence are a small representation of the total volume of information available on the Internet. When researching these topics on the Internet, the chosen Web sites are a starting point for quality, authoritative content. With the arrival of new treatments for impotence and male infertility, there will likely be even more Web sites devoted not only to men's sexual health, but men's health issues in general.

REFERENCES

1. Laumann, Edward O.; Paik, Anthony; and Rosen, Raymond C. "Sexual Dysfunction in the United States: Prevalence and Predictors." *JAMA* 281(February 10, 1999): 537-44.

2. Levine, Laurence A. "Diagnosis and Treatment of Erectile Dysfunction." *American Journal of Medicine* 109(December 18, 2000): 3S-12S.

3. Sadovsky, Richard, and Dunn, Marian E. "Patient-Physician Communication on Men's Sexual Health." Filed at 9:25 a.m. MST, July 23, 2001. Available: <http://urology.medscape.com/Medscape/PrimaryCare/ TreatmentUpdate/1999/tu01/pnt-tu01.html>.

4. "Infertility Myths and Facts." RESOLVE: The National Infertility Association. Filed at 9:30 p.m. MST, July 16, 2001. Available: <http://www.resolve.org/mythfact.htm>.

5. Melman, Arnold, and Gingell, J. Clive. "The Epidemiology and Pathophysiology of Erectile Dysfunction." *Journal of Urology* 161(January 1999): 5-11.

Health Information for Gay Men on the Internet

Gerald (Jerry) Perry

SUMMARY. The groundbreaking publication in April 2001 of the *Healthy People 2010 Companion Document for Lesbian, Gay, Bisexual and Transgender Health* supplement to the nation's *Healthy People 2010* prevention agenda marked a crucial milestone for gay men, as well as other sexual minorities in the United States. Based on work done by the Gay and Lesbian Medical Association and scholars at Columbia University, this policy document outlined the most pressing health concerns for Lesbian, Gay, Bisexual and Transgender (LGBT) people. This article identifies those health concerns specific to gay men, and provides a review of highly select Web resources targeting this population. *[Article copies available for a fee from The Haworth Document Delivery Service: 1-800-HAWORTH E-mail address: <getinfo@haworthpressinc.com> Website: <http://www.HaworthPress.com> © 2002 by The Haworth Press, Inc. All rights reserved.]*

KEYWORDS. Gay men, health information, Internet

Gerald (Jerry) Perry, MLS, AHIP (jperry@ahsl.arizona.edu) is Head of Information Services, Arizona Health Sciences Library, The University of Arizona, 1501 North Campbell Avenue, Tucson, AZ 85724. In the late 1980s and early 1990s, Jerry co-edited three editions of the *AIDS Information Sourcebook* (Phoenix, Oryx Press). He later developed a continuing education course on HIV/AIDS for the Medical Library Association (MLA), and has since published articles on gay men's health and HIV/AIDS information resources in journals such as *Health Care on the Internet*. He is active in the Medical Library Association and was Co-Convener of the Lesbian/Gay/Bisexual/Transgender Health Sciences Librarians Special Interest Group of the MLA.

A longer version of this paper appears in *Internet Reference Services Quarterly*, Vol. 6, No. 2, 2001.

[Haworth co-indexing entry note]: "Health Information for Gay Men on the Internet." Perry, Gerald (Jerry). Co-published simultaneously in *Health Care on the Internet* (The Haworth Information Press, an imprint of The Haworth Press, Inc.) Vol. 6, No. 1/2, 2002, pp. 47-55; and: *Men's Health on the Internet* (ed: M. Sandra Wood, and Janet M. Coggan) The Haworth Information Press, an imprint of The Haworth Press, Inc., 2002, pp. 47-55. Single or multiple copies of this article are available for a fee from The Haworth Document Delivery Service [1-800-HAWORTH, 9:00 a.m. - 5:00 p.m. (EST). E-mail address: getinfo@haworthpressinc.com].

INTRODUCTION

In April 2001, the Gay and Lesbian Medical Association (GLMA) and the National Coalition for Lesbian, Gay, Bisexual and Transgender Health, with funding from the Health Research Services Administration (HRSA) of the U.S. Department of Health and Human Services, published a landmark public policy document on the health concerns for Lesbian, Gay, Bisexual and Transgender (LGBT) citizens. HRSA is responsible for directing national programs to improve the nation's health through assuring equitable access to quality health care for all. The publication, with federal support, of the *Healthy People 2010 Companion Document for Lesbian, Gay, Bisexual and Transgender Health* <http://www.glma.org/policy/hp2010/index.html> supplement to the nation's public policy prevention agenda, *Healthy People 2010* <http://www.health.gov/healthypeople/document/>, marked a crucial moment in the public lives of LBGT people.

To appreciate the publication and what it represents, it is important to understand the *Healthy People 2010* prevention agenda. *Health People 2010* is the United States federal government's illness prevention agenda for all citizens. According to its authors, federal program planners and health care policymakers, "It is designed to serve as a roadmap for improving the health of all people in the United States during the first decade of the 21st century" <http://www.health.gov/healthypeople>. *Healthy People 2010* follows the successful *Healthy People 2000: National Health Promotion and Disease Prevention Objectives* <http://odphp.osophs.dhhs.gov/pubs/hp2000/> initiative, which was released in 1990 and outlined a ten-year strategy for improving the nation's health. *Healthy People 2010* continues that agenda but with a specific focus on addressing underserved and underrepresented citizens, for the most part minorities and socio-economically disadvantaged people. What is startling and groundbreaking about the *Companion Document* is that, for the first time, members of the LGBT community are among the groups recognized by federal policymakers as systematically experiencing health disparities.

CAUSES FOR HEALTH DISPARITIES FOR GAY MEN

The *Healthy People 2010 Companion Document for LGBT Health* was based on scholarship originally conducted jointly by the Gay and

Lesbian Medical Association (GLMA) and the Columbia University Joseph L. Mailman School of Public Health's Center for Gay, Bisexual and Transgender Health. Their research was published in 2000 in the White Paper article, "Lesbian, Gay, Bisexual, and Transgender Health: Findings and Concerns," in the *Journal of the Gay and Lesbian Medical Association* <http://www.glma.org/policy/whitepaper/index.html>.[1] GLMA describes the White Paper as follows:

> This report discusses the health of lesbian, gay male, bisexual and transgender (LGBT) individuals. This grouping includes diverse and varied populations that often share little more than society's stigma and prejudice. Stigma, however, as well as a range of other social and cultural factors, are forces that impact both the health of LGBT people and the ability of health care providers to care for them in myriad ways.[1]

According to the White Paper,

> Homophobia and heterosexism play a role in the inadequate assessment, treatment and prevention of lesbian, gay, and bisexual health problems. LGB individuals suffer from discrimination in housing, employment, and basic civil rights. Homosexuality is still criminalized through sodomy laws in 16 states of the United States, which is [sic] used to deny them jobs, child custody, or participation in the political process . . . Insurance companies, government, hospitals and health clinics often fail to recognize committed lesbian and gay relationships, and deny gay and lesbian partners the privileges granted to married heterosexual couples.[1]

These negative social conditions impact the overall health status of LGBT people.

The White Paper's authors categorize these negative impacts into two broad categories. The first includes the direct consequences of prejudice, such as violence, stress, and poor access to health care services. The second category includes an overall failure in terms of funded research and service policies to adequately address the specific health concerns of LGBT people.

It is in the realm of health services delivery that the disparities experienced by LGBT people are most easily recognized. According to the White Paper's authors,

> Because of negative attitudes prevalent in the U.S. public as well as among physicians and other medical staff, LGBT individuals are subject to discrimination and bias in medical encounters. Moreover, they are likely to receive substandard care, or remain silent about important health issues they fear may lead to stigmatization.[1]

HEALTH PROBLEMS UNIQUE TO GAY MEN

The GLMA/Columbia University White Paper identifies a wide range of health concerns for gay men. Within the realm of sexual behaviors, concerns include HIV/AIDS, hepatitis A and B, enteritis (inflammation of the intestines), and anal cancer. Culturally influenced concerns include eating disorders related to the body and self-image, and alcohol, tobacco, and drug misuse. The health consequences of disclosing one's sexual orientation include depression, anxiety, suicide, and conflicts with family members and friends. Consequences of prejudice and discrimination include harassment from health care providers, limited access to health care insurance, and injuries from anti-gay violence. The health consequences of concealing one's gay identity include reluctance to seek care, delay of treatment, and incomplete or inaccurate medical histories.

QUALITY INFORMATION ABOUT GAY MEN'S HEALTH ON THE INTERNET

If recent studies, such as those conducted by the Cyber Dialogue, Inc. Internet customer relations management company <http://www.cyberdialogue.com/> are correct, nearly half of all adult Americans online are searching the Internet for health information.[2] It is reasonable to assume that among them are gay men.

Gay men will find that the Internet offers a wide range of health information resources sponsored by the government, professional societies, and educational, commercial, and community based organizations (CBOs), all working to promote illness prevention and healthy lifestyles. Additionally, the Net provides gay men with access to information and resources free of the stigmatization they might experience if interacting in person with health care providers or with health information services such as those available at a local public library branch.

And while the scourge of HIV/AIDS continues to impact the gay men's community, Internet information providers have broadened their coverage to include many other potential health concerns of gay men.

Following are annotated references to a select collection of exceptional Web sites about gay men's health. Due to the highly changeable nature of the Internet, if any of the following hypertext links should become inaccessible, a search using a standard Internet search engine such as Google <http://www.google.com> should help uncover the Web site, perhaps at a new location.

Finding the Right Doctor

Gay and Lesbian Medical Association
<http://www.glma.org/>

Championing the health and health care rights of gay men are professional associations, including the Gay and Lesbian Medical Association. GLMA has been instrumental in promoting positive change concerning LGBT health matters at the national level. The association's Web site presents resources and tools primarily intended for its membership. For the average consumer, however, their Web site importantly supports a "Find a GLBT Friendly Doctor" database search feature that allows the gay male user to locate a gay-friendly doctor in his community. Finding the right doctor who will provide care without prejudice is crucial for all people, but for minorities especially.

Finding the Right Information

Tom Flemming's Gay and Lesbian Health
<http://www-hsl.mcmaster.ca/tomflem/gay.htm> and
Gay and Lesbian Health Problems
<http://www-hsl.mcmaster.ca/tomflem/gayprob.html>

Health sciences librarians Tom Flemming, McMaster University, and Jean Sullivant, formerly of the Mount Sinai Medical Center, co-authored a useful article on "Consumer Health Materials for Lesbians, Gay Men, Bisexual and Transgendered People" in the journal, *Public Library Quarterly*.[3] Their article highlights quality health information resources geared to consumers in a variety of formats, including Web sites, and is highly recommended reading. While the article's primary readership is public librarians, consumers will find it a helpful guide.

Tom Flemming is also the author of a well-respected series of Web sites that provide a handy index to reviewed health content on the Web. His *Health Care Information Resources: Illness* and *Wellness* pages provide extensive lists of links to international health-oriented Web content, including resources on *Gay and Lesbian Health Problems* and *Gay and Lesbian Health.*

Typical links from Flemming's sites include those to organizations, (e.g., Gay Lesbian Association of Retired Persons), fact sheets (e.g., Who Do I Tell I Like Other Guys?), personal narratives (e.g., A Coming Out Guide for Gaydads), and Web link collections targeting specific illnesses/conditions (e.g., Becoming Nicotine Free for GLBT Smokers).

Community Based Organization-Sponsored Content

Callen-Lorde Community Health Center <http://www.callen-lorde.org/>

In the 1980s, LGBT CBOs were on the forefront in delivering HIV/AIDS health care information targeted to gay men. Since then, the scope for many health-related LGBT CBOs has expanded, to include other health concerns but also to include the Internet as a forum for disseminating information. The Web site for one high-profile LGBT CBO, New York City's Callen-Lorde Community Health Center, provides an insight into "the nation's premier medical facility for the lesbian, gay, bisexual and transgender communities–as well as people living with HIV/AIDS." The Callen-Lorde Center provides primary medical care: physical exams, immunizations, mental health counseling, and senior wellness services. Their Web site's "Health Information on the Internet" segment provides access to select consumer-oriented health information links on "Cancer," "Cardiovascular Health," "Dermatology," "Health Insurance," "Medication," "Nutrition," "Sexually Transmitted Diseases," "Substance Use," and "Testicular Cancer," among other topics. The site also provides descriptions of the Center's uniquely focused services and mission.

Seattle and King County Public Health's Gay Lesbian Bisexual Transgender Health Webpages <http://www.metrokc.gov/health/glbt/>

LGBT community-focused Web sites issued by state and federal agencies are unfortunately rare. An exception is the exemplary Seattle and King County Public Health's Gay Lesbian Bisexual Transgender Health Webpages. The site's focus is local, with a strong emphasis on regional resources and services to Seattle's LGBT community. The site's "Gay and Bisexual Men" segment features links to information on "Alcohol," "Anal Cancer," "Depression and Mental Health," "Domestic Violence," "Ex-Gay Conversion Therapy," "Finding a Health Care Provider," "Hepatitis," and "Safety and Hate Crimes," among others. Each topic includes a discussion of the health concern as it relates to gay men; some feature recommendations for actions, questionnaires, "pop quizzes," and statistical information and links to regional and national services when available. Overall, and despite its focus on Seattle and the surrounding region, the site is inventive, engaging, and positive, and is highly recommended to anyone interested in gay men's health. What is true for gay men living in Seattle, in terms of specific health problems, is likely to be true for gay men living in other parts of the country.

Commercial Services

GayWellness.com
<http://www.gaywellness.com/>

Health information targeting gay men has commercial value in the new dot.com economy. Numerous sites have arisen, funded with ad dollars, providing health "infotainment" to gay readers. Two examples of this are GayHealth.com, which takes a very high-energy and "newsy" approach to gay men's health concerns, and GayWellness.com, which is significantly less flamboyant though probably as effective in delivering its messages.

GayWellness.com is sponsored by Core Wellness, which also provides a mirror LesbianWellness.com site. According to the introductory comments on the GayWellness home page,

> GayWellness and LesbianWellness focus on the under-served communities of gay men and gay women to provide critically needed education in all areas of wellness such as health maintenance, illness prevention and medical treatment. This knowledge empowers lesbian and gay patients to play an active and informed role in their own well-being through a more complete understanding of their personal wellness issues.

The site's authors describe themselves as

> . . . a team of dedicated healthcare professionals and Internet specialists committed to creating an inclusive learning environment for gay men, lesbians, bisexuals and the transgendered. We believe that wellness is not just the absence of disease, but is the realization of one's physical, emotional, social and spiritual potential.

Content on the GayWellness.com site is organized in a typical magazine issue format, with introductory commentary on each topic accompanied by links to more specific information. For example, the segment, "Day One: The FIRST Step" includes a brief description of the "coming out" process with sidebar links to "Being Gay," and "Meeting Others." The segment on STD's (Sexually Transmitted Diseases) provides an overview with sidebar links to specific diseases such as gonorrhea, genital warts, and herpes, as well as a handy "Symptom Sorter" that provides a list of common signs and symptoms with matching STD.

GayHealth.com
<http://www.gayhealth.com/>

GayHealth.com, on the other hand, exudes a cheerful and chatty appearance, with playful cartoons and images, headline teasers, and lots of opportunities to answer online surveys and cast votes, such as "Viagra: Vote Now!" and "Take the Oral Sex Survey." The Web site looks like the newsy opening pages for a popular gay infotainment magazine such as *Out* or *Curve*. The home page is divided into four columns, like a tabloid newspaper. Quick links occupy the first column, e.g., "News," "Sex," "Emotions," "Food and Fitness," among others. News, online chat forums on sexy topics, and an opportunity to submit a health question to a licensed doctor appear in the third column. Large banner-type ads are featured in the fourth column.

The second column, however, is where the feature articles are. The link on "Sex" for mid-June 2001 featured an article on "Better Sex and Five Other Benefits of Yoga." The link on "Emotions" featured the topics, "Dating and Relating," "Stress and Depression," and "Eating Disorders." It is clear GayHealth.com is interested in integrating gay-specific topics into a broader consideration of an individual's health, and this "whole person" approach is refreshing.

CONCLUSION

The ability to use the Internet to locate and identify with a virtual community of people with shared values has been well demonstrated. For LGBT people, the Internet has provided a platform for community building in light of the ongoing oppression of gay people. Historically, an element of that oppression has been the denial of access to information–information that validates one's life experiences. Like other minorities, gay people are creating the change, whether through the deliberate policy actions of the Gay and Lesbian Medical Association or the commercially sponsored social activism of GayHealth.com.

In the diverse landscape of the Internet, quality gay men's health information is easy to access, available in a variety of formats geared to many different learning styles, and is refreshingly similar to most health information offered up in cyberspace to other men. If "On the Internet, nobody knows you're a dog,"[4] then no one knows that you might be gay, either. Freedom to locate health information without the stigma of labels is the real groundbreaking news at the start of the 21st century for gay men.

REFERENCES

1. Dean, L., Meyer, I. H., Robinson, K. et. al. "Lesbian, Gay, Bisexual, and Transgender Health: Findings and Concerns." *Journal of the Gay and Lesbian Medical Association* 4(3, 2000): 101-51.

2. "Cyber Dialogue Releases Cybercitizen Health 2000," Cyber Dialogue, Inc. <http://www.cyberdialogue.com/news/releases/2000/08-22-cch-launch.html>.

3. Flemming, T., and Sullivant, J. "Consumer Health Materials for Lesbians, Gay Men, Bisexual and Transgendered People." *Public Library Quarterly* 18(3/4, 2000): 95-107.

4. Steiner, P. [cartoon.] *The New Yorker* 69(July 5, 1993).

Men's Health: Sports and Exercise on the Internet

David Howse

SUMMARY. On both a general and sport-specific level, there are many Internet sites that can provide both the novice and advanced male athlete with an abundance of valuable information. Recent hospital emergency room statistics reveal that sports injuries are increasing dramatically across adult age groups, and this coincides with an explosion in new health and fitness related Web sites during the same period. Furthermore, studies suggest that public interest in visiting health and fitness Web sites has been correspondingly high. The author reviews several general and sport specific Web sites, covering areas such as general training and technique, diet, as well as rehabilitation of sports injuries. *[Article copies available for a fee from The Haworth Document Delivery Service: 1-800-HAWORTH. E-mail address: <getinfo@haworthpressinc.com> Website: <http://www.HaworthPress.com> © 2002 by The Haworth Press, Inc. All rights reserved.]*

KEYWORDS. Men, health, sports, fitness, running, soccer, ice hockey, Internet

BACKGROUND

The Internet has been a tremendous boon to people seeking information about health related topics. While many of those going online are

---

David Howse (dhowse@ahsl.arizona.edu) is Information Services Librarian, Arizona Health Sciences Library, University of Arizona, 1501 North Campbell Avenue, P.O. Box 245079, Tucson, AZ 85724.

[Haworth co-indexing entry note]: "Men's Health: Sports and Exercise on the Internet." Howse, David. Co-published simultaneously in *Health Care on the Internet* (The Haworth Information Press, an imprint of The Haworth Press, Inc.) Vol. 6, No. 1/2, 2002, pp. 57-67; and: *Men's Health on the Internet* (ed: M. Sandra Wood, and Janet M. Coggan) The Haworth Information Press, an imprint of The Haworth Press, Inc., 2002, pp. 57-67. Single or multiple copies of this article are available for a fee from The Haworth Document Delivery Service [1-800-HAWORTH, 9:00 a.m. - 5:00 p.m. (EST). E-mail address: getinfo@haworthpressinc.com].

looking for disease-specific information, others simply want to improve health, physical fitness, or athletic performance. Just as there are many fitness Web sites out there that are specifically targeted to women, men are likewise well represented on the Internet at all levels of interest. While the novice fitness buff might want to know how to begin the process of getting in shape, or how to improve eating habits, the intermediate or advanced athlete might be searching for information about how to improve athletic performance. Men's health Web sites are abundant, and herein we shall focus on those that address fitness and exercise generally, while attention will also be given to several popular sport-specific sites as well.

Statistics reveal that many online information seekers are spending a considerable portion of their online time perusing health-related Web sites. A recent study suggests that 52 million American adults, or 55% of those with Internet access, have used the Web to get medical or health information. Furthermore, a striking number are visiting sites concerned with fitness and exercise. For example, according to the same study, 13% of those with Internet access have gone online at some point specifically seeking fitness and nutrition information.[1] Other studies and polls have yielded similar data; a recent poll estimated that 70 million Americans went online between June 1998 and June 1999 to look for information on various health-related topics.[2] It is likely that the subjects of physical fitness and exercise constitute an important subset of these searches. Furthermore, sports injury statistics suggest that a significant percentage of the people who peruse online health resources probably represent the most influential demographic of America society: the baby-boomers.

Health and fitness Web sites are certainly not in short supply, and naturally many are commercial in nature. For example, a typical Web search on the subject of "fitness" will yield search results with sites attempting to sell everything from exercise equipment to health supplements. But like most hobbies and interests covered on the Internet, one can usually come up with a few content-rich gems through persistent digging. Whether you are an established athlete or budding fitness enthusiast, the best sites will cover topics ranging from training or coaching tips, maximizing athletic performance, to common sports injuries.

Sports injuries are a concern for all who enjoy a physically active lifestyle. The Internet has been especially helpful to the consumer interested in finding ways to avoid, or rehabilitate, the most common injuries associated with a specific sport. In addition to being able to find advice on how to avoid aggravating an old sports injury, the information superhighway has made it possible for the athlete to find coaching guidance in cyberspace as well. For example, where once a knowledgeable coach or trainer provided information on such topics as dieting and warm-up exercises, today the athlete can seek out some of this information independently, making it possible to attain higher goals through self-learning. Taking the idea a step further, some Web sites even provide access to online fitness programs, which include coaching service from highly qualified professional trainers, who will monitor your progress and provide advice for a fee. On the other hand, if the athlete sustains an injury, or has a persistent nagging problem from a past injury, there are Web sites that provide information on how to both minimize the discomfort and maximize one's rehabilitation potential.

For the male sports enthusiast, sustaining an injury can almost be considered an inevitable part of participating, especially if one competes or participates over a long term in a physically demanding sport, or one which involves body contact. Statistically, injury rates are considerably higher among young males than females. In a recent study of visits to the emergency room for sports related injuries, it was found that the visit rate for young male patients was more than double the rate for young females.[3]

Sports injury rates among older populations are increasing as well. The baby boomer demographic, for example, has become an increasing presence in hospital emergency rooms over the past few years. Sports-related injuries among those ages 35 to 54–today's baby boomers–increased about 33% from 1991 to 1998.[4] This trend appears to continue into older generations, as statistics show that seniors are also frequenting hospital emergency rooms much more often for treatment for sports related injuries. To illustrate, for those 65 years of age and older, the number of sports-related injuries has gone from 34,000 to 53,000 in the period from 1990 to 1996, an increase of 54% in just a seven-year period.[5] Again, the majority of these older Americans treated for sports injuries were male. It is possible, even likely, that this increase in sports injuries is attributable to people adopting increasingly active lifestyles, including sports participation.

Interestingly, this apparent dramatic increase in sports participation in the 1990s has coincided with the explosion of consumer health information on the Internet. Aside from simply being able to find information more easily than ever before, due to the undeniable convenience of content-rich Web sites, the athlete can also interact with others through bulletin boards or chat rooms devoted to their common interest. For example, the soccer player who sprains an ankle will typically have the injury treated in the emergency department, with rehabilitation advice from the physician or specialist. Later, the curious athlete, determined to return to his sport as quickly as possible, can go online and educate himself about the physiological specifics of what has occurred, based on the doctor's diagnosis and instructions for recuperating from his injury. Anxious to return to the playing field, the athlete can supplement his knowledge by reading online articles or advice on what must happen next, if the injury is to be rehabilitated quickly. In this way, information available on the Internet, providing it is coming from a qualified source, complements the treatment and guidance given by a health care professional, and helps the injured party learn more about the injury and the road to recovery.

The following is a list of general and sport specific Web sites containing valuable information for the male athlete, although females will undoubtedly find several of the sites useful as well. Since it impossible to cover many sports in a short piece, the author has chosen to limit sport specific coverage to ice hockey, soccer, and running.

SPORTS/EXERCISE WEB SITES

Sports Nutrition from Dietsite.com <http://www.dietsite.com/SportsNutrition/index.htm>

For the serious athlete, proper nutrition is essential, and this site provides some excellent guidelines for those committed fitness buffs with high-energy output. Authored by a registered dietician who was once the team nutritionist for the Los Angeles Rams, Sports Nutrition provides insight into how the right approach to diet can influence an athlete's success, while dispelling myths created by advertisements for "wonder drugs" on the market. The author presents a breakdown of the various nutrients for athletes and explains their relationship to successful athletic performance. The site offers high protein and calorie recipes, as well as guidelines for taking "sports foods," such as sports drinks and energy bars. The content is well organized and interesting, but slightly geared toward the serious athlete wanting to maximize performance.

Men'sHealth.com <http://www.menshealth.com/>

As the title denotes, this Web site for the parent *Men's Health* magazine is completely devoted to the subject of men's health, and predictably commits a large amount of space to fitness and sport issues. Unless you are paying for a subscription, however, enabling you to access the online archives, there really is not that much freely available on the Web site. On the other hand, if you are seeking expert advice, or if communicating in cyberspace works better for your lifestyle than consulting with a qualified fitness instructor at your local gym, subscribing to the Men'sHealth.com Personal Trainer service could possibly work very well. The experts staffing this service and answering your questions are highly qualified. This site is especially recommended to those willing to pay the subscription fee, thereby gaining you access to the much larger vault of archived online magazine issues and electronic services.

Boomer-itis
<http://www.boomer-itis.org>

Resulting from a collaboration between the American Academy of Orthopaedic Surgeons (AAOS) and the American Orthopaedic Society for Sports Medicine (AOSSM), Boomer-itis is a joint public education campaign aimed at reducing the number of sports-related injuries among people born between 1946 and 1964. Neatly designed and concise in its coverage, the site provides information on age-related changes, common sports injuries, exercises, and injury prevention tips. The information available is tailored to the area of the body you are interested in. By clicking on the knee of the anatomical image on the home page, for example, you will receive information on the anatomy of the knee, as well as topical links to common knee injuries. For the athlete falling in this age demographic, this is a highly recommended bookmark.

American Council on Exercise (ACE)
<http://www.acefitness.org/>

Although its main purpose is to set national certification and education standards for fitness instructors, the American Council on Exercise (ACE) states that its mission is also to "promote active, healthy lifestyles and their positive effects on the mind, body and spirit." As part of its public education mandate, ACE offers a selection of informative "Fit Facts" and Health "E-tips," available from the main menu of the site's index page. ACE also asserts its role as a "Workout Watchdog," stating, "we strive to protect the public. ACE researches, tests, and challenges fads in exercise products, techniques and trends, and debunks the ad claims of questionable products." For those interested in joining an

ACE endorsed health club, or contacting a local ACE certified trainer, there is a database that can be searched by city, state, or area code. An internationally recognized organization, ACE has a simply organized Web site that provides credible, scientifically based content.

American Orthopaedic Society for Sports Medicine <http://www.sportsmed.org/>

The best features of the AOSSM site are the Patient Education information sheets and the "Ask the Doctor" section, which lists questions neatly by subject area. The problem is that, although the site has excellent content on orthopedic sports injuries, the information does not jump out at you as it should from the main page, but is buried down in the bottom right corner under "Publications and Products." Once found, the organization of topics is very well done, however. Clicking on the "Publications and Products" link on the main page will take you to a "Patient Education" page, containing an excellent "Sports Tips" section on a variety of sports injuries. Tips cover injury definitions, treatment, and rehabilitation recommendations, and can be downloaded in MS-Word format. The "Ask the Sports Doctor" section is organized by topic and contains answers to questions presumably posed by site visitors. With interesting and helpful content for the injured athlete, hopefully this Question/Answer section will be further developed in time with more topics.

American Academy of Orthopaedic Surgeons <http://orthoinfo.aaos.org/>

Similar to the American Orthopaedic Society for Sports Medicine site, the AAOS Web site provides consumer health oriented links to orthopedic conditions and injuries, including sports-related problems. However, this site is better organized for the health information consumer, with links to anatomical joints on the main page taking you to sections containing information about relevant diseases or conditions. Also, a drop-down menu will take you to a "Sports" section, where there is an impressive list of fact sheets and patient education brochures covering injuries typically sustained in various sports.

American College of Foot and Ankle Surgeons <http://www.acfas.org/>

The American College of Foot and Ankle Surgeons is the professional organization for podiatric surgeons. The ACFAS Web site has an excellent Patient Information section, which provides detailed, illustrated brochures on common sports injuries of the foot and ankle, including symptoms and treatment information.

MayoClinic.com–Healthy Living Centers
<http://www.mayoclinic.com/>

The Web site home page for MayoClinic.com is cluttered with links, but seek out the section called "Healthy Living Centers," which is located on the right side of the main page. From the drop-down menu you will see links to "Men's Health" and "Fitness & Sports Medicine." Both provide exceptional content in the areas they cover. Just in case the subject that interests the reader cannot immediately be seen, there is a search engine available which covers the entire Web site. It is worth using, because, as is often the case in medicine, subject matter is often spread across several categories. There are detailed illustrations which are appropriately placed, and a variety of "personal planners" for exercise, stress, and weight management. Depending on the subject matter you are reading, relevant news and related topics are placed strategically on the page for easy and intuitive navigation. For those with a broadband connection, there is an abundance of multimedia available. Confused about proper form for the bicep curl when weight training? Download a short video with instructions. This site is an excellent resource for all athletes.

The Physician and Sportsmedicine Online
<http://www.physsportsmed.com/index.html>

The Physician and Sportsmedicine is a peer-reviewed monthly journal serving the practicing physician's professional and personal interests in the medical aspects of exercise, sports, and fitness. Surprisingly, most of the content in the current and back issues are available online freely from the journal's Web site. In addition to providing access to the full-text journal articles, with an accompanying search engine, the site provides excellent consumer health oriented information via the "Personal Health" and "Resource Center" sections. The former contains first-rate articles written by physicians on a variety of sports medicine topics, while the latter provides relevant directories and links for the sports medicine community. Even if the full-text access to the online journal is eventually withdrawn, there is still some valuable content available here.

MEDLINEplus–Exercise/Physical Fitness <http://www.nlm.nih.gov/medlineplus/exercisephysicalfitness.html>

Setting the standard for consumer health information, the National Library of Medicine's MEDLINEplus must be included in this list. "Exercise/Physical Fitness" is just one of many health topics covered in this database of links to mostly external, dependable Web-based sources. Organized under commonly used broad categories such as "General/ Overviews" or "Specific Conditions/Aspects," the site will provide the fitness enthusiast with information originating from organizations generally associated with medicine, rather than linking to sports or fitness oriented organizations.

Male Health Center <http://www.malehealthcenter.com/>

Taking a broader view of men's health, this site provides some interesting content from the syndicated newspaper columnist on men's health issues, urologist Ken Goldberg. Goldberg, who contributes to the advisory panels for National Men's Health Week, Men's Health Network and *Men's Health Magazine,* provides links to a number of his columns on a variety of subjects, including nutrition and exercise. The site is easy to navigate; the columns are written in the first person and generally intended to motivate the male reader to engage in healthy activities.

Runner's World: Health and Fitness <http://www.runnersworld.com./healthy/home.html>

Highly regarded by those who take running seriously, *Runner's World* magazine provides a well-organized, comprehensive Web site that goes beyond simply providing free information. There is an emphasis on emotional and motivational support for the beginning and experienced runner, and the site covers such issues as stretching, cross-training, weight loss, injuries, staying motivated, and nutrition. A beginner's program covers a variety of important issues the novice should consider, including everything from shoe selection, to how to react to sore muscles, to diet recommendations. There is an extensive section on food and nutrition, with a list of recipes for those concerned with maintaining a proper diet for running optimally. Those seeking to communicate with, or seek out advice from, other runners, can register to use the public forums as well.

American Running Association
<http://www.americanrunning.org/>

The American Running Association was founded "by running doctors to help more people benefit from a running lifestyle." There is an abundance of information available to the nonmember, including educational brochures and journal articles covering a wide range of running-related subjects available in PDF or HTML format. Additional services are available to members. Membership gets you a subscription to the *Running and FitNews* newsletter, free medical training/advice, answers to fitness questions, use of the running shoe database, and free maps to running trails located throughout the U.S.

Soccer Performance.org
< http://www.soccerperformance.org/>

Soccer Performance.org is a personal Web site. The author approaches the subject academically and has compiled information from personal research, taken from "books, the internet or that [which] has been presented to me by knowledgeable sources, or from my own experiences as a soccer player, student and coach." Despite the fact that this Web site is not expressly affiliated with, or endorsed by, soccer governing bodies or organizations, the information content is first-rate, with articles from top-notch contributing authors, some of whom have affiliations with world-class professional clubs. Well-designed, the site features some very helpful articles on tactics, performance, and fitness issues, all specifically tailored to the soccer player. Complementing the text are impressive instructions, complete with diagrams and, in some cases, animations. This site is a wonderful resource for serious students of the sport of soccer, from the perspective of both coach and player.

Finesoccer.com
<http://finesoccer.com/>

Soccer is one of the fastest growing sports in North America. Finesoccer.com was created to help soccer coaches and players improve their knowledge of the sport, and to get the most out of their training. The visitor is encouraged to subscribe to a series of free, e-mailed newsletters, and the site features an online archive of drills and tactical knowledge useful to players and coaches.

US Youth Soccer Online
<http://www.youthsoccer.org/>

US Youth Soccer is a non-profit and educational organization whose mission is to foster the physical, mental, and emotional growth and development of America's youth through the sport of soccer at all levels of age and competition. Despite its focus on youth, this site has separate archives of "Training Tips" and "Health Tips," which benefit players of all ages.

Hockeyinjuries.com
<http://www.hockeyinjuries.com/>

Hockeyinjuries.com is dedicated to providing players, parents, coaches, and trainers with the most comprehensive information related to the prevention and management of common ice hockey injuries. The staff writers include the team physician, as well as the team physical therapist, of the National Hockey League's Calgary Flames. Considering the legal implications of the subject matter, there is a clever legal disclaimer built in. Before you can choose any of the links to the site's areas, you must first read a disclaimer indicating that the site's content is provided for informational purposes only, and is intended to supplement, not substitute for, the advice of a physician. Very easily navigated and endorsed by the Canadian Academy of Sport Medicine, the content covers areas such as injury care and prevention and nutrition, but also includes interesting features such as an "Injury Calculator" and an "Equipment Quiz." The former is intended to provide advice about how long after sustaining a specific injury it will typically take to return to play, while the latter assesses hockey equipment safety. For general injury information, the site visitor can click part of a hockey player's body, and then be transported to a list of typical hockey injuries affecting that area, complete with anatomical and color photographs or illustrations where appropriate. This site is an excellent resource for hockey players of all ages.

Pashby Sport Concussion Safety Web Site
<http://www.concussionsafety.com/>

Concussions are a serious concern for athletes in physical contact sports. This site provides an overview of this serious injury, including statistics, as well as short videos showing how concussions typically occur, and what happens physiologically to the head and brain during impact. Additionally, there is information about how the concussion grading system works, how to prevent and recognize concussions, as well as instructions on how the athlete should return to competition (i.e., only under medical supervision). The information is clearly designed to educate the public about an injury that occurs frequently in sports such as boxing, ice hockey and football. A Canadian site with an ice hockey focus, it is easily navigated and concise in content.

REFERENCES

1. Fox, Suzannah, and Rainie, Lee. "The Online Health Care Revolution: How the Web Helps Americans Take Better Care of Themselves." Pew Internet and American Life Project. 2000. Available: <http://www.pewinternet.org>. Accessed: July 9, 2001.

2. The Harris Poll. "Explosive Growth of Cyberchontriacs Continues." August 5, 1999.

3. Burt, Catherine W., and Overpeck, Mary D. "Emergency Visits for Sports-Related Injuries." *Annals of Emergency Medicine* 37(March 2001):301-8.

4. U.S. Consumer Product Safety Commission. "Baby Boomer Sports Injuries." Available: <http://www.cpsc.gov/LIBRARY/boomer.pdf>. April 2000. Accessed: July 9, 2001.

5. U.S. Consumer Product Safety Commission. "Sports Related Injuries to Persons 65 Years of Age and Older." Available: <http://www.cpsc.gov/CPSCPUB/PUBS/GRAND/aging/injury65.PDF>. April 1998. Accessed: July 9, 2001.

The Silent Killer

Jack W. Raines

SUMMARY. High blood pressure or hypertension continues as a major medical problem in the United States and the world. Hypertension leads to stroke, heart attack, and other coronary diseases. Generally, except for those past age 70, it is a much greater problem for males than for females. Half of those with high blood pressure are not even aware of it and for over 90%, the cause is unknown. Blood pressure, and the resulting coronary problems, can be reduced significantly with lifestyle and health choices. The information presented at most of the medical Web sites applies to both men and women. Many of the behavioral changes are relatively simple to accomplish, but men need to be aware of those healthy lifestyle choices if they are to reduce their risks of later coronary disease. *[Article copies available for a fee from The Haworth Document Delivery Service: 1-800-HAWORTH. E-mail address: <getinfo@haworthpressinc.com> Website: <http://www.HaworthPress.com> © 2002 by The Haworth Press, Inc. All rights reserved.]*

KEYWORDS. Men's health, high blood pressure, hypertension, stroke, heart attack, coronary disease, coronary heart disease, CHD

INTRODUCTION

More than 61 million people in the United States have high blood pressure, or hypertension, and nearly half do not know they have it, ac-

---

Jack W. Raines, MAT, MLIS (jraines@hsc.unt.edu) is Extension Librarian, Gibson D. Lewis Library, 3500 Camp Bowie Boulevard, University of North Texas Health Science Center, Fort Worth, TX 76107-2699.

[Haworth co-indexing entry note]: "The Silent Killer." Raines, Jack W. Co-published simultaneously in *Health Care on the Internet* (The Haworth Information Press, an imprint of The Haworth Press, Inc.) Vol. 6, No. 1/2, 2002, pp. 69-78; and: *Men's Health on the Internet* (ed: M. Sandra Wood, and Janet M. Coggan) The Haworth Information Press, an imprint of The Haworth Press, Inc., 2002, pp. 69-78. Single or multiple copies of this article are available for a fee from The Haworth Document Delivery Service [1-800-HAWORTH, 9:00 a.m. - 5:00 p.m. (EST). E-mail address: getinfo@haworthpressinc.com].

cording to the American Heart Association.[1] Hypertension can lead to stroke, heart attack, and other coronary diseases and, because of this, hypertension is often referred to as the "silent killer."[2]

There are many organizations worldwide that collect statistics on heart diseases. Comparing statistics can be extremely valuable, and those who analyze the statistics can present important information to a targeted group. The National Center for Health Statistics has a Web section called "fast stats A to Z." The section contains a link to a large statistical table showing hypertension among persons 20 years of age and over, according to sex, age, race, and Hispanic origin for the United States, comparing four time blocks from 1960 to 1994.[3-4] From the statistics given, it is possible to conclude that a bigger problem with hypertension exists as men age, and that hypertension is of greater statistical significance to black males. But it is probably not necessary to compare much beyond simple numbers to reach an equally valid conclusion that, regardless of age or race, hypertension remains a serious problem in men's health, affecting a very large number of men.

Hypertension is defined as a disorder characterized by high blood pressure where generally, systolic blood pressure (the "top" number, the pressure generated when the heart beats) is consistently higher than 140, and diastolic blood pressure (the "bottom" number, the pressure in the vessels when the heart is at rest) is consistently over 90.[5] Blood pressure is determined by the amount of blood pumped by the heart, and the size and condition of the arteries. Many factors can affect blood pressure, including the volume of water in the body; the salt content of the body; condition of the kidneys, nervous system, or blood vessels; and levels of various hormones in the body.

In 90% to 95% of high blood pressure cases, the cause is unknown and a person can have high blood pressure for years without knowing it.[6] For this reason, checking with a physician and being tested is essential. The information presented in this paper is designed to augment the information given by medical professionals as it applies to the individual patient. Even when one is diagnosed, since the cause is most often unknown, treatments are primarily for control of the condition rather than curing the condition. Much information is available on the World Wide Web for patients as well as medical professionals concerning all aspects of control, including diet and exercise, and a general healthy lifestyle, as well as medications needed for some cases.

The Internet sites reviewed in this article have been evaluated as to their quality including currency, accuracy, and a lack of bias. This is not an exhaustive list. Any general search tool will yield many sites with information about hypertension. It is the responsibility of the person conducting the search to evaluate the information; any ideas for alterations to general lifestyles, such as diet and exercise programs, should not be undertaken without first consulting one's physician.

ORGANIZATIONAL REFERENCES

American Heart Association <http://www.americanheart.org/>

The American Heart Association (AHA) is the logical first place to look for information concerning hypertension. It is a fairly easy site to negotiate, but with a few problems relative to this specific report. Most of the information presented concerning hypertension and high blood pressure is not gender specific to males. There is a section concerning infant health and women's health, but no section concerning men's health. Very up-to-date statistical information can be found easily at the hyperlinked site for "Heart and Stroke A to Z Guide."[7] The term "hypertension" is cross-referenced to high blood pressure and there are several hyperlinks to general and specific topics concerning hypertension. There is also a separate section concerning high blood pressure, although it is not intuitively clear how to get to that section since it is not linked from the " . . . A to Z Guide." Once any one of the links concerning high blood pressure from that guide is used, there are additional links, and at the end of any of those pages, links back to the "High Blood Pressure" Section.

American Stroke Association <http://www.strokeassociation.org/index.html>

The American Stroke Association is a division of the American Heart Association. There are several links at the site to direct the user to the AHA and to various AHA sponsored programs that cross the divisions of the AHA. The best example of such a link is the One Of A Kind[SM] personalized health management program. It is an interactive health and behavior change program designed for the consumer and patient where specific information can be put into the program with personalized results available. The program discusses smoking, physical inactivity, nutrition, and medication compliance. The idea is to provide a customized program with "information on how to start following a healthier diet, how to increase the physical activity you get and more. And it's all customized to your own needs."[8]

U.S. GOVERNMENT SITES

National Heart, Lung, and Blood Institute <http://www.nhlbi.nih.gov/index.htm>

The National Heart, Lung, and Blood Institute (NHLBI), part of the National Institutes of Health under the Department of Health and Human Services, provides leadership for a national program in diseases of the heart, blood vessels, lung, and blood. Additionally, they sponsor research and education about sleep disorders and have administrative responsibility for the NIH Woman's Health Initiative and the Healthy People 2010 projects.

Information is arranged for patients and the general public in one section and for health care professionals in another section. In the first section, for patients, a general section for heart and vascular diseases appears. There is a section about high blood pressure with excellent information that is both easy to read and very informative. There are many subsections, but approximately half of those concern women's health and there is nothing directly aimed at men. The same situation exists in the other sections concerning related problems of heart attacks and general cardiovascular diseases.

High blood pressure, as previously mentioned, is particularly prevalent in Hispanic and/or African-American males. There are many resources on this site directed to those two specific groups, but none directed specifically to black or Hispanic men or men's health. There are additional reports, available in a downloadable format, many produced by the NHLBI, in the section for health care professionals. These reports are generally directed toward specific racial and ethnic groups and tend to report specific results from NHLBI sponsored research.

National Library of Medicine Gateway <http://gateway.nlm.nih.gov/gw/Cmd>

One of the most complete sources of information concerning any medical condition is the United States National Library of Medicine (NLM). There are several ways to access the information contained through this resource. Those who want to get all the information available in one source can use the NLM Gateway, designed for users to search in multiple databases at the NLM with a single search. The site will give results in MEDLINE, LOCATORplus®, MEDLINEPlus®, DIRLINE, and any other applicable databases produced by the NLM.[9] The Gateway is a work in progress and new databases regularly get added to that interface. The goal is to have eventually most of the NLM databases searchable with one single interface. Changes to the interface are also often made to enhance or refine the searching capabilities and parameters.

A search for hypertension will reveal over 200 thousand items, mostly journal citations from MEDLINE, but also books, serials, audio-visual materials, consumer health articles, meeting abstracts, and others. The lists can be displayed, but most of them have only abstracts or basic bibliographic information available. Also, the subject of hypertension cannot be limited to either sex in the interface search screen. Complicating the use of the NLM Gateway for this search is the fact that "men's health" is not a MeSH term so a general search on that concept cannot be retrieved separately.

MEDLINEplus® Health Topics <http://www.nlm.nih.gov/medlineplus/healthtopics.html>

MEDLINEplus® is the consumer health reference source from the National Library of Medicine. The materials are written for the consumer and articles and information are presented using general terminology rather than medical terminology. Men's health, as used in the listing, is concerned primarily with sexually transmitted diseases, infertility, reproductive health, and prostate diseases. In order to find information concerning hypertension, the easier of the two ways is to go directly to the heart and circulation category. And then, since this index uses lay terminology, the obvious choice is high blood pressure.

The other method that can be used is to select topics beginning with the letter H and choose high blood pressure from the list. Once the topic is chosen, there is a wealth of information concerning hypertension. Note that, as previously mentioned, this is not male gender specific. What will be found are the latest news reports from agencies such as the Associated Press and Reuters. This is particularly valuable if there has been a news report on television, Internet news services, or in the local newspapers.

The remaining information will vary, but generally will cover such areas as overview information, definitions, management, prevention, treatment, the latest research, and links to related organizations. Again, it must be noted that there are sections concerning children, seniors, and women, but not a separate section concerning men. There is also a section, as with most subjects in MEDLINEplus®, of general Spanish language resources, but none of these are gender specific to males.

MEDLINEplus®, under the subject of high blood pressure, contains an interactive module concerning general aspects of hypertension. This is part of the NLM's "X-Plain Online" series from the Patient Education Institute <http://www.nlm.nih.gov/medlineplus/tutorials/hypertension.html>. There are general information screens that discuss definitions of blood pressure measurements, causes, complications, diagnosis, and controls for hypertension; all contain interactive questions to confirm one's understanding.

Also in MEDLINEplus®, under the term stroke, appear the same categories and type of general information as listed above from the National Institutes of Health. "Know Stroke" is from the National Institute of Neurological Disorders and Stroke, one of the many NIH organizations <http://www.ninds.nih.gov/health_and_medical/disorder_index.htm>.

The page for disorders is organized, as are so many of the general NIH sites, so that one can choose the letter "S" and then select "stroke." The page with stroke information has four main areas: definition, treatment, prognosis, and research. There are links to other organizations and related NINDS publications which are full-text online and can be downloaded, read online, or printed.

GENERAL SITE

About.com/Men's Health <http://menshealth.about.com/health/menshealth/index.htm>

Probably one of the best known sites for general information on the Web is About.com. This is a network of approximately 700 guide sites organized into about 40 channels or general subject areas. These cover about 50,000 individual subjects with over one million links. The index started in 1997 as "The Mining Company" and was renamed in 1997 as About.com. Individuals who are usually experts in a particular field compile a list of links to relevant pages of major medical sites throughout the Internet. This procedure eliminates the need to go to a medical site's home page and then search for the desired topic.

Under the main page for men's health a sidebar is displayed listing relevant subjects, including heart health, hypertension, and stroke. A person can select any one of these three subjects and will be given first a definition of the term and a synopsis of relevant subtopics. The next screen shows titles of articles with an abstract and a hyperlink to the article. The full article is opened in a window within the About.com site. This can be extremely confusing for the user because there is not necessarily an easy way to tell where the original article is located on the Web. Often, depending on the original site's design, there is no way to get to the home page of the organization or person who owns the site, making it difficult to evaluate the site for accuracy, timeliness, objectivity, bias, and authority. The claim from the About.com site is that the person who manages their specific page is an authority and has evaluated the site for overall quality.

The second problem is that the links often go to sites on the Web that may or may not be men's health sites. Often, as in the cases of several of the links, they go to excellent articles from major medical journals, such as the *Journal of the American Medical Association,* but are not necessarily gender-specific or even directed to men generally.

UK SITES

Bandolier
<http://www.jr2.ox.ac.uk/bandolier/index.html>

Bandolier is an evidence-based health care site based in the United Kingdom. Under their "specialist subsites" there are several references of interest. The general men's health section is primarily concerned with prostate cancer, benign prostatic hyperplasia, and sexual health. There is a section designated "cardiac," which discusses hypertension, strokes, and heart attacks in reports concerning specific categories of diets, food supplements, and behaviors, as well as heart disease treatments and management. The site has a search capability on the home page, and titles are clearly indexed and in general categories that are easy to understand. The index is available for the current issue of the site with new health related stories in each monthly issue. Previous issues are available and articles have been included since 1995. The "healthy living" section introduction states: "These healthy living pages are being developed from medical literature. Associations are examined mainly between behaviours and health outcomes, such as reducing the risk of hypertension as a result of taking exercise."[10]

Health Technology Assessment Page
<http://www.hta.nhsweb.nhs.uk/>

The National Coordinating Centre for Health Technology Assessment, a division of the United Kingdom's National Health Service, coordinates the HTA Program under contract from the Department of Health's R&D Directorate. The home page has a search form that quickly links to full-text research reports from various medical and health organizations and universities in the UK. Searching for various terms, such as high blood pressure, hypertension, stroke, or heart disease, will result in approximately one hundred reports, giving both the medical professional and layperson the results of the latest research. Whereas much of the information in the reports is very technical, there is information concerning results of research that will be of interest to patients and their families.

Men's Health Forum
<http://www.menshealthforum.org.uk/>

Malehealth
<http://www.malehealth.co.uk/>

Two related sites are The Men's Health Forum, primarily for medical professionals, and Malehealth, claiming to be "the definitive site for anyone who wants to know more about the health of men and boys."[11] Both sites are produced by Radcliffe Online, the Web information site for Radcliffe Medical Publishers in the United Kingdom. Men's Health Forum has many statistical reports and evaluations or summaries of government documents relating to health issues, many concerned primarily with the UK itself. However, from the home page of the Malehealth site, there is a section called "What's Up Doc?" containing advice and information about a wide range of common health problems, including coronary heart disease (CHD) which covers high blood pressure and stroke. It is an extremely well organized site in that many additional terms used in the discussion are further defined and explained, such as "blood pressure" and "diabetes." The discussion is aimed at males specifically, and in their discussion of obesity, they refer to obesity at the waist as age increases. Of course, the site uses contemporary and common British terminology, such as "big belly obesity," but all of the terms are generally intuitive to Americans as they are used in the text.

In addition to the "What's Up Doc?" section, there are nine other sections dealing with news, self testing, print references, fitness, a chat room for health issues, and general health information. It is an excellent site, especially since the discussion is totally directed toward men's health.

ALTERNATIVE MEDICINE

HealthWorld Online <http://www.healthy.net/index.asp>

HealthWorld Online advertises itself as " the only Internet health network that integrates both alternative and conventional health information into a synergistic whole."[12] The site is well accepted by general Internet evaluations, such as Lycos and Yahoo, from the technology and design view. It is a very large site with a few hundred individual pages and scores of internal links. There is a section called "healthy man," and as with most of the sites previously mentioned, most of it refers to male sexual health. There are, however, additional subjects, called "men's conditions," not generally found in the other male health sites, including baldness, athlete's foot, testicular cancer, and sports injuries. However, the subjects of hypertension, heart disease, and stroke are all individual topics listed under "health conditions" and are not gender specific.

CONCLUSION

There are obviously many Internet sites available concerning high blood pressure and all resulting medical problems and conditions. A large number of Internet sites devoted to the issue of men's health currently exist in the development phase. However, there are very few that discuss hypertension as specifically related to men. Hypertension continues as a major health concern for men, with more than 50% of men in certain age groups having high blood pressure. But it is a problem that is also significant for the whole population, and most of the behaviors that aggravate the condition are the same for both sexes. Men need to be aware of those life style choices that need to be changed to reduce the many dangers associated with hypertension.

REFERENCES

1. Cook, Allan R. *Men's Health Concerns Sourcebook: Basic Information about Health*. Detroit: Omnigraphics, 1998, 54.

2. American Heart Association. "High Blood Pressure." Available: <http://www.americanheart.org/hbp/index.jsp>. Accessed: June 28, 2001.

3. National Center for Health Statistics. "faststats a to z." Available: <http://www.cdc.gov/nchs/fastats/>. Accessed: June 28, 2001.

4. National Center for Health Statistics. Statistical Tables, Charts, and Graphs. "Table 66. Hypertension Among Persons 20 years of Age and Over, According to Sex, Age, Race and Hispanic Origin: United States, 1960-62, 1971-74, 1976-80, and 1988-94." Available: <http://www.cdc.gov/nchs/products/pubs/pubd/hus/tables/2000/00hus066.pdf >. Accessed: June 28, 2001.

5. MEDLINEplus®, Dictionaries. Adam Medical Encyclopedia. Hypertension. "Definition." Available: <http://www.nlm.nih.gov/medlineplus/ency/article/000468.htm#contentDescription>. Accessed: June 28, 2001.

6. American Heart Association. About Blood Pressure. "What Causes High Blood Pressure?" Available: <http://www.americanheart.org/hbp/about_causes.html>. Accessed: June 28, 2001.

7. American Heart Association. Heart and Stroke A-Z Guide. Cross-Reference Index. "2001 Statistical Update." Available: <http://www.americanheart.org/Heart_and_Stroke_A_Z_Guide/>. Accessed: June 28, 2001.

8. American Heart Association. You're One of a Kind and So is Your Life. "What Kind of Information Will I Receive?" Available: <http://www.onelife.americanheart.org/welcome/info1.cfm#information>. Accessed: June 28, 2001.

9. NLM. "Gateway." Available: <http://gateway.nlm.nih.gov/gw/Cmd?GMBasicSearch>. Accessed: June 28, 2001.

10. Bandolier. "Healthy Living." Available: <http://www.jr2.ox.ac.uk/bandolier/booth/booths/hliving.html>. Accessed: June 28, 2001.

11. MALEHEALTH. "Men's Health FORUM." Available: <http://www.malehealth.co.uk>. Accessed: June 28, 2001.

12. HealthWorld Online. Available: <http://www.healthy.net/index.asp>. Accessed: June 28, 2001.

Internet Information on Heart Disease in Men

Richard Nollan

SUMMARY. Finding information on heart disease in men might seem as easy as logging on to a favorite search engine, finding the right category, and perusing the available links. However, in trying several prominent search engines, this turned out not to be the case. One important reason for the lack of information is that heart disease has declined 60% since 1950. Nevertheless, heart disease is still the leading killer among men. An advanced search on a search engine (heart disease as a phrase and man or men) still yields an unwieldy number of pages, about 90,000. In an effort to bridge this gap, the author examines a number of governmental, organizational, and consumer Web sites containing high-quality information on heart disease. *[Article copies available for a fee from The Haworth Document Delivery Service: 1-800-HAWORTH. E-mail address: <getinfo@haworthpressinc.com> Website: <http://www.HaworthPress.com> © 2002 by The Haworth Press, Inc. All rights reserved.]*

KEYWORDS. Internet, heart disease, search engines, consumer health Web sites

Richard Nollan (rnollan@utmem.edu) is Special Collections Librarian and Reference Librarian, University of Tennessee Health Science Center Library, 877 Madison Avenue, Memphis, TN 38163-0001. He has served continuously in reference since 1987, including ten years as the Clinical Branch Librarian. Mr. Nollan manages three Web sites.

[Haworth co-indexing entry note]: "Internet Information on Heart Disease in Men." Nollan, Richard. Co-published simultaneously in *Health Care on the Internet* (The Haworth Information Press, an imprint of The Haworth Press, Inc.) Vol. 6, No. 1/2, 2002, pp. 79-89; and: *Men's Health on the Internet* (ed: M. Sandra Wood, and Janet M. Coggan) The Haworth Information Press, an imprint of The Haworth Press, Inc., 2002, pp. 79-89. Single or multiple copies of this article are available for a fee from The Haworth Document Delivery Service [1-800-HAWORTH, 9:00 a.m. - 5:00 p.m. (EST). E-mail address: getinfo@haworthpressinc.com].

A quick glance at Internet search engines reveals that under the broad category of men's health, few sites have a category for heart disease. Not long ago, heart disease in men was a popular topic, generating much interest among physicians and concern among patients. At sites such as Yahoo and Alta Vista, one can find categories such as benign prostatic hyperplasia, body dysmorphic disorder, circumcision, male breast cancer, and priapism, but none on heart disease. The reason for the absence of heart disease on many men's health sites is undoubtedly the dramatic decline of the disease in recent decades. The Centers for Disease Control in Atlanta reported in 1999 that cardiovascular disease has dropped by 60% since 1950. Factors contributing to this decline include public health campaigns, clinical trials, a decline in smoking, better control of blood pressure, decreased cholesterol levels, and improved treatments. While this constitutes a striking success for modern medicine, the lack of access to relevant information can leave those seeking it at something of a loss.

Finding dedicated sites with high quality information can be challenging. There are relatively few of them. Although a search engine will yield hundreds of pages of links, the vast majority of them appear to be stories by news agencies, drug companies, or other organizations whose main purpose is something other than communicating health care information. Examples of dedicated sites listed below include the National Heart, Lung, and Blood Institute and the American Heart Association. The sites offering comprehensive health care information contain sections on heart disease that are as good as the dedicated sites, including NOAH, WebMD, and ClinWEB International.

BACKGROUND

Despite the decline in heart disease, however, there is still cause for concern. Cardiovascular disease represents the leading cause of death among men. In 1998, nearly half a million men died of heart disease, a number equal to the sum of the next four leading causes.[1] Men are at greater risk from heart disease if they smoke, have a diet rich in animal fats that is not balanced with whole food, are related to someone who died of it, have high blood pressure and high cholesterol levels, do not exercise, and are over the age of forty. More than one-third of the men who die between the ages of 45 and 65 die of a heart attack. It is also true that, at any age, heart disease occurs more often in men than in women. Warning signs of a heart attack include sudden crushing pain in the chest, palpitations, shortness of breath, and profuse sweating. Treatments range from prescription drugs to open up the vessels to major heart bypass operations.

All sites listed in this article were chosen because they are created by reputable organizations rather than by a single author. They have a proven track record in that the sponsors have successfully organized health care information for many years, such as the National Heart, Lung, and Blood Institute, or the heart is their major focus, such as the American Heart Association. All sites make some disclaimer about how their information should be used or about the completeness of their information.

Since these sites often ask users to register (usually at no cost, such as for information tailored to a user's medical condition or for a newsletter), they also contain a statement about privacy. This is particularly important for those searching for information on their medical condition, but who don't want such personal information shared with others. The privacy concern is one that is growing increasingly important as a national medical issue. Some commercial sites use the registration process to gather names that can later be sold to other companies as distribution lists.

Each of the following sites contains current information. However, currency is measured in different ways. One popular conception holds information to be current if the ink is still wet on the page. No site professes this kind of currency. Some Web sites try to accommodate this desire by presenting news flashes with links to an appropriate story, or index links that are organized by categories, for example heart disease broken down into arrhythmia, valvular, and other disorders. If currency is important, the reader must look for it where and how the site creators provide it. A reputable site will indicate how recent its information is.

GOVERNMENT WEB SITES

In the short history of the Internet, government Web sites rank with educational Web sites as the most stable sites containing the most reliable information. The federal government distributes a kaleidoscopic array of support for research, all of which must be made publicly available. Much free information on heart disease is available in digital format. Reports hundreds of pages long can be downloaded in a matter of minutes and then searched or printed.

The National Heart, Lung, and Blood Institute <http://www.nhlbi.nih.gov/index.htm>

NATIONAL INSTITUTES OF HEALTH
National Heart, Lung, and Blood Institute

The Institute plans, conducts, fosters, and supports an integrated and coordinated program of basic research, clinical investigations and trials, observational studies, and demonstration and education projects. Research is related to the causes, prevention, diagnosis, and treatment of heart, blood vessel, lung, and blood diseases and sleep disorders. The NHLBI plans and directs research in development and evaluation of interventions and devices related to prevention, treatment, and rehabilitation of patients suffering from such diseases and disorders. It also supports research on clinical use of blood and all aspects of the management of blood resources. Research is conducted in the Institute's own laboratories and by scientific institutions and individuals supported by research grants and contracts.

Centers for Disease Control and Prevention <http://www.cdc.gov>

Located in Atlanta, Georgia, the Centers for Disease Control and Prevention (CDC) is an agency of the Department of Health and Human Services that promotes health and quality of life by preventing and controlling disease, injury, and disability. The CDC monitors health, detects and investigates health problems, conducts research to enhance prevention, develops and advocates sound public health policies, implements prevention strategies, promotes healthy behaviors, fosters safe and healthful environments, and provides leadership and training. Its Web site offers a storehouse of statistical data on the range of diseases in the United States. The list of CDC publications includes the *MMWR (Morbidity and Mortality Weekly Report)* and *Emerging Infectious Diseases*. The *Wonder* database is searchable on all the statistics and reports contained on the site. Under "Health Topics A-Z" is a list of disorders, including heart disease, where the user will find a publication *Men and Heart Disease,* a book-length report containing a wealth of helpful data. The site is available in both English and Spanish.

ORGANIZATIONAL WEB SITES

The American Heart Association <http://www.americanheart.org>

While the National Heart, Lung, and Blood Institute is a clearinghouse for information on all kinds of medical issues, the American Heart Association site is dedicated solely to heart disease and stroke. This site is invaluable for both the professional and the consumer. In addition to general topics, such as warning signs, heart and stroke A-Z guide, family health, and risk awareness, the site offers drop-down subject menus for the professional and the consumer. There is an abundance of current information on the causes and treatment of heart disease, as well as statistical resources on all aspects of heart disease and stroke.

Heart Information Network: heartinfo.org <http://www.heartinfo.org/>

HeartInfo is an independent, educational Web site dedicated to providing a wide range of information and services to heart patients and to those interested in learning about lowering their risk factors for heart disease. Their database includes a wide range of articles and news stories. The site offers little background information on the founders and little about the organization itself.

CONSUMER WEB SITES

These sites are chosen for their reliability and stability, but also because they offer quality information aimed at helping the consumer understand what is important in the prevention, diagnosis, treatment, and rehabilitation of heart disease.

About: Heart Disease/Cardiology
<http://heartdisease.about.com/mbody.htm>

This site offers a wealth of information related to heart health and disease in the form of articles, forums, chats, categories and a newsletter. The organizer, or "guide," of this site is Richard N. Fogoros, M.D., cardiologist. The topics here are numerous and detailed, and range from angioplasty/stents and aortic aneurysm to syncope and valvular disease. The organizers of this site are attempting to pull everything they can find related to the heart into this site.

Illness: Health Care Information Resources. Heart Disease Links
<http://www-hsl.mcmaster.ca/tomflem/heartdis.html>

This page contains links to about 100 sites with information on heart disease and related illnesses or to sites with information about organizations in this area. The links go to American and Canadian sites, and, depending on the locations, the information they contain will either be in English or Spanish. It is produced at the McMaster University Health Sciences Library in Hamilton, Ontario.

WebMD Health
<http://my.webmd.com/index>

This commercial site has a broad range of information on both the diseased and the healthy heart. The editors gather information from a wide array of sources, including news services, and in some cases, create articles as they are needed. WebMD Health offers assistance in the form of information, research, and support, such as discussion lists. Registration is free, but allows the user to create a profile of just the kind of information that he or she is seeking. The site also provides e-mail links for asking questions of a physician.

CliniWEB International
<http://www.ohsu.edu/cliniweb/>

CliniWeb is a list of links to clinical information on the Internet and is maintained by the Oregon Health and Science University. The links pertaining to heart disease are extensive and organized by disease type. Although its search engine, called SAPPHIRE, can be operated in five languages (English, German, French, Spanish, Portuguese), this feature is limited by the fact that most of the site's information is in English. This comprehensive site contains nearly 10,000 selected URL's, but makes no claim to including everything on the Web. Where completeness is important, the site contains links to other search engines. However, for the experienced searcher it is worth noting that the site is organized according to the National Library of Medicine's Medical Subject Headings (MeSH).

Hardin MD: Heart Disease and Cardiology
<http://www.lib.uiowa.edu/hardin/md/cardio.html>

This list of heart sites is maintained by the University of Iowa's Hardin Library for the Health Sciences. It claims to contain only the "best" sites, which it determines by the percentage of working links on its pages. If a list site has less than 93% actively working links on its site, then the site is considered to be poorly maintained, and it is not included in Hardin MD. It also ranks sites by the number of links they contain, namely large, medium, and small, and the site can be searched by keyword.

HealthWeb: Cardiology
<http://healthweb.org/>

This site is the product of a collaboration of over twenty actively participating health science libraries in the Midwest. The libraries' work is supported by a contract between the National Library of Medicine and the University of Illinois in Chicago. The site's cardiology section contains a list of links to a range of Web sites by professional and other organizations. It consists of annotated consumer health resources. The site also offers the option of doing a keyword search on the entire site, but a search that is too specific (e.g., arrhythmia) may overlook useful sites. Categories include consumer health resources, educational resources, and health care professionals. There is no clear statement in the description of the site as to how sites are selected, evaluated, or monitored.

MayoClinic.com: Heart & Blood Vessels Center <http://www.mayohealth.org/>

This consumer health site is distinguished by being an extension of the public health history of the world famous Mayo Clinic. It is created by a team of editors, health care professionals, and media producers, and is owned by Mayo Foundation for Medical Education and Research. The site is reevaluated annually. The site focuses on both health and disease in broad categories, and it can be searched by keyword.

Medical Matrix <http://www.medmatrix.org>

Medical Matrix is a categorized and annotated list of links focusing on clinical medicine. Selected sites are reviewed by its editorial board, which is drawn from the American Medical Informatics Association's Internet Working Group. Each review results in assigning from one to five stars, with one star being "suitable, well-authored clinical content but lacking in substance, or currency." Five stars refer to "an award-winning site for Medical Internet." Despite its clinical orientation, the heart disease section contains consumer health information. The site contains links to full-text journal articles and books, as well as to multimedia sites, continuing education, prescription assistance, clinical searches, and news. In addition to categories, the site can also be searched by keyword.

MedWeb <http://www.medweb.emory.edu/MedWeb/>

MedWeb is a list of biomedical and health related Web sites maintained by the staff of the Robert W. Woodruff Health Sciences Center Library at Emory University. Its primary audience is the academic and research community at Emory, but some attention is paid to the consumer's search for health information. Its guidelines for evaluating sites include relevancy to Emory's needs, credibility, currency, content, and design. Although not a comprehensive site *per se,* the institution and its clearly delineated criteria are broad and this lends their list of links some authority. The list of cardiology links numbers nearly 300; these link to Web sites and to publications on heart related topics.

familydoctor.org: Health Information for the Whole Family from the American Academy of Family Physicians <http://familydoctor.org/>

The American Academy of Family Physicians is a national organization of family physicians, who regularly review and update the information in this site. Categories of information on this site include handouts, fact sheets, drug information (including drug interactions), herbal and alternative remedies, a directory, and information on self-care. There are categories for heart disease and also one on how the cardiovascular system works. It contains a section just for men, which, among other subjects, discusses heart disease and related disorders, including assessing and reducing the risk of heart disease, and warning signs for and rehabilitation from heart attack. The site is in English, with many items available in Spanish.

InteliHealth <http://www.intelihealth.com/>

The medical information used on the site comes from a variety of sources, including the Harvard Medical School and the University of Pennsylvania. Drug information is derived from the United States Pharmacopoeia Drug Information (USPDI) and the Johns Hopkins Drug Encyclopedia. The site has a separate men's heart and circulatory section, with information about heart disease and how the cardiovascular system works. Articles are updated every two years, with the most recent update indicated in the article. There is a search window for the site, and an "Ask the doc" e-mail link. InteliHealth is a commercial site that focuses on providing "the most credible health information on the Internet" to consumers. The site displays advertising, but also clearly states in its editorial policy that none of the information providers used on the site endorses any of the advertised products or the information produced for the site. An editorial board maintains InteliHealth.

The Merck Manual of Diagnosis and Therapy, 17th Edition <http://www.merck.com/pubs/mmanual/>

Published by Merck and Company, this venerable textbook has been on doctor's shelves for over a century. Because of its reputation as a high-quality general text, it has found a place among consumer texts. The site has a search window, but also allows for searching by table of contents. This online version contains a section on Cardiovascular Disorders, with seventeen chapters on the range of heart disorders. Because it is easier to incorporate changes and updates online, the online version varies slightly from the paper version.

NOAH: New York Online Access to Health
<http://www.noah-health.org/>

New York Online Access to Health (NOAH) is the product of four organizations: The City University of New York, The Metropolitan New York Library Council, The New York Academy of Medicine, and The New York Public Library. This comprehensive collection of links in both English and Spanish has been the recipient of numerous awards for content and ease of use. Links on heart disease are distributed over three broad categories, including care and treatment, information resources, and "What is heart disease and stroke?"

PDR.net
<http://www.pdr.net>

The *Physicians Desk Reference* (*PDR*) is another highly respected source of drug information for physicians that has existed for over one hundred years. The site includes access to a number of other resources as well, including full text of the *PDR, PDR Herbal Medicine,* the full text to selected journals and magazines, patient education, and *Stedman's Medical Dictionary*. Registration is required. Although this site does not have a cardiology section, it can be useful for those looking for information on specific drugs used to treat heart disease. The free service includes access to everything except the *PDR* and *Stedman's*. Access to the latter two can be paid for in a monthly or annual fee. Although the *PDR* focuses on drug information, heart disease often is prominently featured, from aspirin to tissue plasminogen activator.

CONCLUSION

Although no single site can offer a truly comprehensive collection of information sources or even links to those sources, the selected sites in this paper come as close to that ideal as possible. Nevertheless, it's worth remembering how quickly conditions on the Internet can change.

Even the most reliable sites can change their policy or disappear altogether. What may be available this week, may cost money in the future, may be replaced, or may be removed. With so much information available on heart disease, it remains necessary for the searcher to persist and to be resourceful in searching the Internet.

REFERENCE

1. Centers for Disease Control. "Achievements in Public Health, 1900-1999: Decline in Death from Heart Disease and Stroke–United States, 1900-1999." *MMWR* 48(August 6, 1999):649-56.

Male Menopause–Myth or Mystery?

Patrice Hall
Valerie Gross

SUMMARY. As they age, men face physiological changes similar to what women experience during menopause. This article reviews Web sites that discuss "male menopause" and its suitability as a diagnosis. It also reviews more general sites that provide information on healthy aging for men. *[Article copies available for a fee from The Haworth Document Delivery Service: 1-800-HAWORTH. E-mail address: <getinfo@haworthpressinc.com> Website: <http://www.HaworthPress.com> © 2002 by The Haworth Press, Inc. All rights reserved.]*

KEYWORDS. Male menopause, andropause, Internet

INTRODUCTION

Men are notorious for not taking care of their health. A national survey showed that 1 in 10, or approximately 7 million men have avoided getting regular health exams for more than a decade. Overall, slightly

---

Patrice Hall (pmhall@psu.edu) is Librarian, Lois High Berstler Community Health Library, The Milton S. Hershey Medical Center, P.O. Box 850 HS-07, Hershey, PA 17033-0850. Valerie Gross (vgross@geisinger.edu) is Librarian, Community Health Resource Library, Geisinger Medical Center, 100 North Academy Drive, Danville, PA 17822-2020.

[Haworth co-indexing entry note]: "Male Menopause–Myth or Mystery?" Hall, Patrice and Valerie Gross. Co-published simultaneously in *Health Care on the Internet* (The Haworth Information Press, an imprint of The Haworth Press, Inc.) Vol. 6, No. 1/2, 2002, pp. 91-98; and: *Men's Health on the Internet* (ed: M. Sandra Wood, and Janet M. Coggan) The Haworth Information Press, an imprint of The Haworth Press, Inc., 2002, pp. 91-98. Single or multiple copies of this article are available for a fee from The Haworth Document Delivery Service [1-800-HAWORTH, 9:00 a.m. - 5:00 p.m. (EST). E-mail address: getinfo@haworthpressinc.com].

more than 15 million men have not had a basic health check in five years or more.[1] Although Web resources are never a substitute for good medical care, there are sites that provide information on the changes men will face as they age and on how to improve their chances for a healthier and longer life.

ANDROPAUSE/MALE MENOPAUSE

A particular health problem many men will face as they age is "male menopause." Its symptoms include erectile dysfunction, urinary problems, mood swings, irritability, and depression. But is male menopause or andropause, as it is sometimes called, a real disease? All men experience a loss of testosterone as they age. Should this change be considered a disease and treated with hormone replacement? Or is it simply a natural part of aging with which men need to cope? The medical community has been debating these questions for a number of years and there are Web sites that cover both sides of this issue.

MayoClinic.com
<http://www.mayoclinic.com/home?id=MC00003>

The MayoClinic.com site contains an excellent article on male menopause. It discusses the difference between a midlife crisis and andropause. There is a link to an article on testosterone replacement. This vetted site is a good place to begin researching male menopause.

Andrology
<http://www.andrology.com/>

This site is sponsored by the Andromeda Andrology Center in Hyderabad, India. Male menopause, impotence, infertility, and Peyronie's Disease are some of the topics covered. The graphics are clear and helpful. This is an informative site but no sources are listed.

Andropause: Canadian Andropause Society
<http://www.andropause.com>

The information provided at this site is general and brief. It covers the condition, causes, impact, and options. There are no graphics. There are no sources listed, and no editorial board is listed.

Andropause Canada
<http://www.andropausecanada.com>

Another Canadian site, it is part of LifeMD.com. The editorial board consists of health care professionals, including physicians and nurses. Educational grants and sponsors provide the funding for this site. The site conforms to the HON code of conduct principles. It includes information about when andropause occurs, symptoms, short and long-term effects, and treatment. The treatment section mentions only injectable or oral testosterone therapy. Non-drug regimes such as improved diet, exercise, and a reduction of stress are not mentioned.

The Andropause Society
<http://www.andropause.org.uk>

The Andropause Society is a charity based in the United Kingdom. Its mission is the promotion and exchange of information and ideas about andropause. Offered here is information about the history, causes, treatment, and effects of andropause. Members can receive greater full-text information; however, there is a membership fee. Sources are not listed.

Male Health Centres
<http://www.malehealth.com/HTML/andropause.html>

Male Health Centres, based in Canada, specialize in erectile dysfunction and prostate disease. The Centre lists the clinical trials in which it participates. The site also has a section about andropause. It provides information about the condition itself, the importance of testosterone, the effects of andropause, and treatment options. Sources are not listed nor is an editorial board.

Midlife Passages
<http://www.midlife-passages.com/page25.html>

This site is the child of Andrew B. Dott, M.D., M.P.H., F.A.C.O.G., Obstetrics and Gynecology and Caroline Dott, M.S.W., Ph.D., Clinical Psychology and Research. There is a section on andropause that discusses the symptoms, causes, risks, and treatments. The treatment section addresses hormone replacement therapy for men, and the different methods of administering them (pills, transdermal preparations, injections, and subdermal pellets). Suggestions for further reading are made.

About.com Health & Fitness

The Health & Fitness channel of About.Com has a number of Guide Sites that contain links to articles on andropause and other health issues for the middle-aged male. Each About.com site has a subject expert who is responsible for selecting links, writing original articles, and moderating message boards and chats. Original content and links on the sites are good. Also, links to informational sites and sponsor sites are clearly distinguished. However, an excess of ads, including those most annoying pop-up ads, detract from the sites.

About.com Women's Health
<http://womenshealth.about.com/health/womenshealth/library/blandropause.htm>

Offers a good overview of andropause and relates the pros and cons of hormone replacement therapy. Also available are suggestions for further reading and links to other sites containing information about andropause.

About.com Men's Health
<http://menshealth.about.com/health/menshealth/cs/midlifecrisis>

This section contains links to sites that discuss whether or not andropause should be considered a real condition.

About.com Senior Health
<http://www.seniorhealth.about.com/health/seniorhealth/cs/menshealth/>

This Guide Site includes links to sites on topics such as benign prostate hyperplasia, impotence, gout, male breast cancer, and hormone replacement for men.

AGING AND MEN'S HEALTH

Loss of libido is not the only health concern of the aging male. According to the CDC, the ten leading causes of death in men are heart disease, cancer, unintentional injuries, cerebrovascular diseases, COPD, pneumonia and influenza, diabetes mellitus, suicide, chronic liver disease and cirrhosis, and homicide.[2] Choosing a healthier lifestyle can prevent many of these diseases. In addition to sites that focus on andropause and midlife crisis, there are many Web sites that address these other health concerns of the aging male.

AARP
<http://www.aarp.org>

The American Association of Retired Persons Web site has extensive information available on issues of interest to the aging male. Its Wellness Center has articles on topics such as staying active, nutrition, and stress. There are links to a variety of Web resources and a recommended reading list. Other sections of interest include Life Transitions and Money & Work. The site is uncluttered, easy to navigate, and content-rich.

AgeNet
<http://www.agenet.com>

AgeNet is a commercial site that provides information and referral services to older adults and their caregivers. Its interactive tools include an Eldercare Locator, which has a searchable list of senior centers, hospices, attorneys, and geriatricians; a CareGiver Tools section, which contains checklists on nursing homes, safety, and medications; and an Ask the Experts section. The Health section contains news updates and articles on topics such as depression, diabetes, stroke, chronic pain, and Parkinson's disease. This site would be useful to any man who finds himself part of the "sandwich generation" and is trying to take care of his own health and that of his parents.

Alliance for Aging Research
<http://www.agingresearch.org>

Founded in 1986, the Alliance for Aging Research is a non-profit organization dedicated to improving the health and independence of Americans as they age by advocating and promoting greater scientific and medical research in aging. Its Web site includes information on its activities and the latest news on aging research and government legislative and policy changes that impact on the "boomer" generation. The Health Topics section contains a browsable list that includes age-related macular degeneration, diet and nutrition, cholesterol and heart disease, regenerative medicine, and end of life issues. The newsletter, *Living Longer & Loving It,* requires a free membership.

American Dietetic Association
<http://www.eatright.org/olderamericans>

Part of the Web site of the American Dietetic Association, this site provides information on special dietary needs of the aging. It includes a food guide pyramid for persons 50 plus and fact sheets on antitoxidants, fiber, protein, grains, and calcium.

MEDLINEplus Seniors' Health Topics
<http://www.nlm.nih.gov/medlineplus/seniorshealth.html>

This section of MEDLINEplus provides an extensive list of health topics including arthritis, Alzheimer's disease, impotence, prostatic diseases, and stroke. Selecting a topic takes the user to an authoritative list of resources that includes news articles, clinical trials, organizations, directories, and articles.

Merck Manual of Geriatrics
<http://www.merck.com/pubs/mm_geriatrics/search.html>

Written for the professional, *The Merck Manual of Geriatrics* is designed to provide information to all those involved in the care of elderly patients. Topics covered include nutrition, infectious disease, cardiovascular, pulmonary, gastrointestinal, genitourinary, hematological and psychiatric disorders, and legal, ethical, and social issues. This full-text online version is searchable by keyword and has a browsable table of contents and index.

National Institute on Aging
<http://www.nih.gov/nia/>

Part of the National Institutes of Health, the National Institute on Aging conducts research and disseminates information on the aging process and age-related diseases. Its Web site contains full-text pamphlets and booklets on a variety of topics. *Age Pages,* which are short, informative brochures, include aging and alcohol abuse, sexuality and aging, coping with diabetes, and depression. Longer booklets available on the site include *Exercise: A Guide from the National Institute on Aging, Hearts & Arteries,* and *Talking with Your Doctor: A Guide for Older People.*

RealAge.com
<http://www.realage.com>

RealAge is an interactive site that allows users to calculate their "Real Age" as opposed to their chronological age. One hundred twenty-five different factors comprising medical, genetic, psychological, lifestyle, and environmental risks are considered in the Real Age calculation. Once the user's biological age is determined, a personalized health plan is developed, which recommends lifestyle and health changes. Besides the main questionnaire, there are also Real Age Interactive Guides on specific health conditions. Available topics of interest to the aging male include enlarged prostate, prostate cancer screening, Type 2 Diabetes, cholesterol, and hangovers. The site also has a Medical Encyclopedia that is made up of Clinical Reference Systems: Behavioral Health Advisor, Adult Health Advisor, and Medication Advisor. It can be browsed or searched. The site has a four-star (the highest) rating in Enonymous.com's privacy rating guide and subscribes to the HONcode principles of the Health On the Net Foundation.

Staying Healthy at 50+
<http://www.ahrq.gov/ppip/50plus/index.html>

Published by the Agency for Healthcare Research and Quality, this guide covers healthy lifestyle changes and discusses which screening tests and immunizations older adults need. It includes contact information for many government health institutes and personal health information charts that can be printed out.

ThirdAge.com
<http://www.thirdage.com/>

ThirdAge.com is a commercial site for "Third-agers," who are defined as the first wave of Baby Boomers. The Web site's stated mission is to "develop activity and information-rich communities–where members can assess life choices, master their computers, find value in the World Wide Web, exchange wisdom, and connect with others who share their passions and interests." The health section includes topics such as arthritis, sexuality, nutrition, elder care, and heart health. Information includes a medical encyclopedia from Adam.com, news articles from Associated Press, and articles written by subject experts. Most content is available without login; however, the free membership includes access to e-mail, chats, personals, and discounts. This is a content-rich site, but too many flashing ads detract from its appearance.

CONCLUSION

A recent Survey by the Pew Internet & American Life Project showed that men are more likely to be using their time on the Web looking for sports information or stock quotes rather than health information.[3] Informative Web sites exist that can help ease men's transition into midlife. Men need to be encouraged to expand their use of the Web and explore these resources. Armed with this information they and the people who love them can live healthier, more fulfilling lives.

REFERENCES

1. Illinois Department of Aging. "Facts on Aging . . . About Men's Health." Issue 5, May 15, 1999. Available: <http://www.state.il.us/aging/onage05.htm>.

2. National Center for Health Statistics. Health, United States, 2000 With Adolescent Health Chartbook. Hyattsville, MD: NCHS, 2000. Available: <http://www.cdc.gov/products/pubs/pubd/hus/tables/2000/00hus032.pdf>.

3. "Women Online Catch Up to Men . . . but They Surf Differently." *Fort Worth Business Press*, June 9, 2000, V 13, Issue 7, p. 11A.

Using the Web as a Resource on Hair Loss

Deborah G. Lovett

SUMMARY. Loss of hair (alopecia) is one of the most common symptoms of aging in men. The most common cause of hair loss in men is androgenic or androgenetic alopecia, often referred to as male-pattern baldness. Current treatments include two drugs, Rogaine and Propecia, which have been shown effective for hair regrowth and prevention of further hair loss; surgical hair transplantation; and cosmetic options. Many patients turn to the Internet for health information, and Web sites dedicated to hair loss products and clinics abound. Consumer health sites, pharmaceutical sites, and professional organizational sites dealing with hair loss are reviewed, including sites that are appropriate for physicians and health care providers, as well as consumers. *[Article copies available for a fee from The Haworth Document Delivery Service: 1-800-HAWORTH. E-mail address: <getinfo@haworthpressinc.com> Website: <http://www.HaworthPress.com> © 2002 by The Haworth Press, Inc. All rights reserved.]*

KEYWORDS. Hair loss, alopecia, male pattern baldness, Internet

INTRODUCTION

Loss of hair (alopecia) is one of the most common symptoms of aging in men. While there are many different types of alopecia, the most common is androgenic or androgenetic alopecia, often referred to as

---

Deborah G. Lovett (dlovett@psu.edu) is Assistant Librarian, Educational Services, The George T. Harrell Library, The Pennsylvania State University College of Medicine, The Milton S. Hershey Medical Center, Hershey, PA 17033.

[Haworth co-indexing entry note]: "Using the Web as a Resource on Hair Loss." Lovett, Deborah G. Co-published simultaneously in *Health Care on the Internet* (The Haworth Information Press, an imprint of The Haworth Press, Inc.) Vol. 6, No. 1/2, 2002, pp. 99-109; and: *Men's Health on the Internet* (ed: M. Sandra Wood, and Janet M. Coggan) The Haworth Information Press, an imprint of The Haworth Press, Inc., 2002, pp. 99-109. Single or multiple copies of this article are available for a fee from The Haworth Document Delivery Service [1-800-HAWORTH, 9:00 a.m. - 5:00 p.m. (EST). E-mail address: getinfo@haworthpressinc.com].

"male-pattern baldness." Up to 80% of all men lose their hair to varying degrees, and 95% of that hair loss is due to androgenetic alopecia.[1]

Genetic predisposition and hormonal factors have been identified as the major factors in androgenetic hair loss. Hair grows in two to six-year cycles, with a growth phase, a transitional phase, and a resting phase. Each hair passes through the phases independent of the neighboring hairs. Dihydroxytestosterone (DHT), a byproduct of testosterone metabolism, circulates through the blood stream and causes the hair in genetically predetermined areas of the scalp to become shorter, finer, and less pigmented with each hair growth cycle. This process starts as early as age 20, with most men experiencing some hair loss by the age of 50.[2]

The loss of hair affects one's self-image, causing many men to search for ways to regrow their hair. At this point, baldness cannot be "cured," but there are two drugs, Rogaine (minoxidil) and Propecia (finasteride), which have proven effective for hair regrowth and prevention of further hair loss in some patients. There are currently other drugs under clinical trial that, similar to Propecia, work by suppressing DHT production. Surgical hair transplantation is another form of effective treatment and involves grafting healthy hair follicles from unaffected parts of the scalp to the bald spots. Other options include herbal preparations, many of which are under investigation for effectiveness, and cosmetic hair pieces.

There is a plethora of Internet sites relating to hair loss, many of which are commercial sites for hair loss products and clinics. A Google search on the term "hair loss" generated a list of over 700,000 hits. This article will highlight some of the Internet sites that provide quality information on a variety of issues related to hair loss. Sites that promote one particular product or physician are not included in this discussion, which will focus instead on sites that provide quality information on hair loss, the treatments currently available, and those who provide that treatment.

CONSUMER MEGASITES

It is always appropriate to start an Internet search at the major consumer health sites. With the exception of MEDLINEplus, these sites accept advertising; however, they all provide good quality, non-biased health information.

Helioshealth
<http://www.helioshealth.com>

Helioshealth provides current information on various health conditions that has been reviewed by a medical advisory board comprised of physician specialists. This site includes a general description of male pattern baldness; an explanation of how hair grows; and the causes, diagnosis, and treatment of hair loss.

MayoClinic
<http://www.mayoclinic.com>

Male pattern baldness is discussed on three different pages of this site including a page on alopecia, a page on male hormonal changes, and a page on hair loss. In addition to a general description of baldness and its treatment, there is also information on the signs and symptoms of hair loss, as well as a section on coping strategies.

Intelihealth
<http://www.intelihealth.com>

Several pages discuss hair loss and male pattern baldness. One page gives an overview of all the causes of hair loss, and then several pages specifically discuss hereditary baldness and effective treatments.

MEDLINEplus
<http://www.nlm.nih.gov/medlineplus>

There is an excellent medical encyclopedia on this site which provides information on the various causes of hair loss as well as a section on male pattern baldness which includes illustrations and information on the causes and risks; prevention; symptoms; signs and tests; treatment; prognosis; and complications.

PHARMACEUTICAL SITES

Because there are drugs available for treatment of androgenetic alopecia, searching for information on these drugs is also warranted. Merck, the pharmaceutical company that produces Propecia, has two sites, one for physicians and one for consumers, while the Upjohn site on Rogaine is designed specifically for consumers.

Merck for Physicians
<http://www.merck.com/product/usa/propecia/hcp/home.html>

The physician site includes full prescribing information including description, clinical pharmacology, indications and usage, contraindications, warnings, precautions, adverse reactions, dosage and administration, and how supplied. There is clinical data on the effectiveness of the drug, before and after pictures of patients taking Propecia, and a profile of those who can safely take the drug.

Merck for Consumers
<http://www.keephair.com>

This glitzy site has various sections including:

- Talk About Hair Loss–Short videos of men and women talking about what hair and baldness means to them;
- Male Pattern Hair Loss–Detailed description of the most common cause of hair loss amongst men;
- Famous Hairlines and Unusual Remedies from the Past–A look at famous people throughout the ages with and without hair and a description of the various ways people tried to stop hair loss in the past;
- Treatments and Answers–Survey of treatments available for hair loss in men and answers to frequently asked questions about hair loss and Propecia; and
- Interactive Fun–Activities including a hair loss quiz, flirting with the virtual nurse, and changing your hairline after scanning in a picture.

The site works best with Flash 4 and a high-speed Internet connection, but will work on a regular connection as well. The video talks require Real Video Player software and a link is provided for those who need to download the player.

Rogaine
<http://www.rogaine.com>

The Upjohn site on Rogaine is designed for consumers and has a site for women as well as a site for men. The site provides useful information on how the drug works, the results that can be expected, and FAQ's

(Frequently Asked Questions), in addition to a community forum which includes nine discussion groups on dealing with hair loss, using Rogaine, product support, and testimonials. In the "How It Works" section of the site there is detailed information on the proper application of the product. Upjohn provides a free month's supply of the product for signing up with the "Rogaine Results" e-mail support group.

PROFESSIONAL ORGANIZATION SITES

The effective treatment of alopecia requires consultation with a physician. There are many professional organizations that have created Web sites of interest in this discussion. The sites reviewed here provide consumer information on various aspects of hair restoration. Most of the sites offer a "find a specialist" link. While any physician can treat this condition, it would certainly be prudent for a patient to be sure that the physician has some expertise in the treatment of hair loss. Some dermatologists and surgeons have sub-specialized in hair restoration. Many physicians will state on their Web sites that they are "board certified," and there are several professional organizations that grant certification. However, the issue of certification is a very complex one. In an effort to be sure that physicians are qualified, the American Board of Medical Specialists was created. Various academies and boards were formed by the medical specialists with the purpose of assuring that members have adequate education and expertise to practice in that area of medicine. Once certification is granted, continuing education hours and an examination every few years is required for physicians to maintain certification. At this time, the American Board of Medical Specialists does not recognize any certification for hair restoration specialists. However, many hair restoration specialists are certified in other areas through the American Academy of Dermatology, the American Board of Plastic Surgery, and the American Board of Surgery.

The organizations that claim to certify physicians in hair restoration are: the International Society of Hair Restoration Surgery; the American Society of Hair Restoration Surgery; the World Association for Hair Restoration Surgeons; the American Academy of Facial, Plastic and Reconstructive Surgery; and the American Board of Hair Restoration Surgery. The certification is granted in various ways. Some organizations grant certification based simply on membership in the organization, while others have more stringent guidelines including a minimum number of procedures performed, or an examination. In light of the previous discussion, use the "find a specialist" links found on many of these sites with reservation.

American Academy of Dermatology
<http://www.aad.org>

The American Academy of Dermatology is the largest of all dermatologic associations, with a membership of over 13,700 representing virtually all practicing dermatologists in the United States. Their Web site has a "members only" section as well as a general section. Of interest for this discussion are the patient education and patient information sections of the site. There are patient education pamphlets on many dermatologic problems including hair loss. The copyrighted brochure on hair loss includes an explanation of normal hair growth and the causes of excessive hair loss. In the patient information section there are also various articles on hair loss and baldness, including "Halt Hair Loss . . . Now," "Transplanting in Microspace: Winning the War Against Baldness in the Third Millennium," and "Into Thin Hair." The site also has a "find a specialist" section that links the user to certified dermatologists in a particular geographic area.

American Hair Loss Council
<http://www.ahlc.org>

The AHLC is an organization that provides non-biased information about the options for hair loss treatment. They certify non-physician hair loss specialists while an affiliated organization, the American Board of Hair Restoration Surgery, provides certification for physicians who perform surgical hair restoration. There is a "find a member" link, though this is of limited use since the AHLC is not a large organization. The online HTML patient education brochures are excellent and cover various hair-loss topics, including Hair Replacement Surgery, Non-Surgical Hair Additions, Alopecia Areata, Male Pattern Hair Loss, Female Pattern Hair Loss, and Chemotherapy-Related Hair Loss.

American Society for Dermatologic Surgery
<http://www.asds-net.org/hair_restoration.html>

This organization's Web site presents information on various dermatologic conditions including hair loss, the treatment options, and techniques of hair restoration. The society provides an online, procedure-based, geographic-specific physician referral service of board-certified or board-eligible physicians who perform dermatologic surgery. The referral section is comprised of various drop-down menus and is easy to use.

Canadian Hair Research Foundation
<http://www.hairinfo.org>

The CHRF has designed a comprehensive site providing information for patients and physicians on the various causes and treatments of alopecia. In the Learning Centre, there is historical information on hair loss, a description of current research needs, and downloadable patient education brochures, including "Hair Loss," "Hair Loss in Men-Myths and Facts," "Hair Loss in Women," and "Treatment Options for Hair Loss Due to Androgenetic Alopecia." The foundation has also made available three medical education lectures for physicians on the management of androgenetic alopecia. This site requires the use of RealPlayer and Acrobat Reader for downloads.

International Society of Hair Restoration Surgery
<http://www.ishrs.org>

This organization provides certification for hair restoration surgeons worldwide and has produced an attractive site. Information on both surgical and non-surgical treatments can be found on the site as well as sections on causes, patient experiences, and links to other sites. Of particular noteworthiness is the FAQ section, which provides answers to common questions about hair transplantation surgery such as: "Does it really work?" "What can be expected immediately after surgery?" "Is it painful?" "What will it look like after the procedure?" "When can I go back to work?" The ISHRS provides a "find a doctor" link, but with only 700 members, it is of limited benefit.

The American Academy of Facial Plastic and Reconstructive Surgery
<http://www.aafprs.org/patient>

Some surgeons who perform hair replacement surgery are members of this organization. There is an explanation of the board certification process and general information about the academy on this page and on the "FAQ's" page. The "Procedures" page has an excellent explanation of hair replacement surgery, a glossary of terms, and before and after pictures. There is also a physician finder listing board-certified plastic surgeons who are members of the academy.

Internet Dermatology Society
<http://www.telemedicine.org>

The IDS is comprised of physicians, students, and scientists who are involved in the field of dermatology. The site includes: the "Electronic Textbook of Dermatology," "Dermatology Knowledge Builder," a global overview of Internet dermatology resources, access to "Dermatology Lectures," and "Global Dermatology Grand Rounds." The resources on this site are not designed for consumers, but rather are targeted to clinical practicing dermatologists.

GENERAL CONSUMER SITES

The sites included in this section are authored by people who either have male pattern baldness or work with those who do. All are funded through advertising by pharmaceutical companies and companies offering hair loss products. The sites that have been selected provide non-biased information. However, some of these sites will have links to other sites that may not present unbiased information.

Keratin.com
<http://www.keratin.com>

Written by a research immunologist/dermatologist, this Web site is devoted to educating consumers about hair loss and presents good information on the biology of hair as well as the various types of alopecias, including effluviums, inflammatory alopecias, hair shaft defects, infectious disease, hypertrichosis, and alopecia areata, in addition to androgenetic alopecia. In the "Diagnosis/Decisions" section of the Web site there are pages about spotting a hair loss scam, questions to ask before choosing a hair loss treatment, what you can expect when visiting a dermatologist, and how a diagnosis is made. The section on "Hair Cosmetics" provides various non-medical options for disguising hair loss including permanent cosmetics as well as a discussion on hair care.

Hair Transplant Network <http://www.hairtransplantnetwork.com>

This site was "created by and for hair transplant patients," and most of the site is dedicated to explaining micro-follicular unit hair transplantation surgery. There are pictures documenting how the hair transplants are done as well as before and after pictures. Though there is no overt advertising on this site, several hair loss clinics/surgeons are promoted on the "find a surgeon" page. However, there is an excellent section on considerations in evaluating hair transplant clinics with questions that patients should ask, such as the surgeon's reputation among peers, availability of "before and after" case photos, number of procedures performed, and where they are performed.

The Regrowth Network <http://www.regrowth.com>

The Regrowth Network mission is "to provide individuals suffering from hair loss with the most complete, accurate, and timely information on hair loss prevention, reversal and other solutions, so that they may make an educated and confident decision about how to treat their problem." This Web site only began accepting advertising recently, and the founder is committed to keeping editorial content separate from advertising.

The site is comprised of the following sections:

- Recent–information on current research on alopecia;
- Reactive–online chat, discussion groups, and surveys;
- Reference–information on treatments, transplants, hair systems, conferences and female hair loss;
- Reviews–book and product reviews including most of the major treatments by name,
- Results–photos and information on hair regrowth; and
- Registration–form for registering with the site owner.

The best part of this site is the Reference section on the page labeled "treatments," in which you can search for current information on various "Medicinal," "Natural," "Commercial," and "Future" treatments. This information has been organized in a tabular format that includes the drug type, availability, claimed and observed results, clinical results, safety/side effects, testing, and typical cost. Links to MEDLINE references are provided for each treatment discussed in this section.

Follicle.com
<http://www.follicle.com>

This site covers the most common types of hair loss, but has a special focus on androgenetic alopecia. Pages within the site are: "Hair Structure and Life Cycle," "Common Types of Hair Loss," "Hair Loss Treatments," a glossary, and links. Under "Hair Loss Treatments" there are two categories of treatments discussed: medical treatments for androgenetic alopecia and natural treatments for androgenetic alopecia. Clicking on a treatment option brings up an explanation of how the treatment works, its effectiveness, side effects, and studies associated with the treatment, including the reference.

One concern with this site is that there is no indication of who is responsible for the site content. Despite this major drawback, it has good information presented in an eye-pleasing format.

HairLoss Advisor
<http://www.hairlossadvisor.com>

The goal of the HairLoss Advisor is to provide information and sound advice on hair loss prevention and treatment. This site is run by Healthology, an online health media company owned and managed by physicians, and includes general information on hair loss as well as featured articles on the cosmetic options for hair loss and surgical options for hair loss. There are four video presentations available that require the use of RealVideo. They are: "Choosing a Hair Loss Expert," "The Future of Hair Loss Treatment," "Pills & Creams: Medical Treatments for Hair Loss in Men," and "Can Your Diet Help You Keep Your Hair?"

Scams
<http://www.hairlossscams.com>

This site is dedicated to protecting people from the various hair loss scams that are abundant on the Internet and is sponsored by the producers of a cosmetic product, DermMatch. Because of the glut of commercial sites for hair loss treatments available for purchase on the Internet, this site is very useful in providing guidelines to help consumers identify scams from legitimate treatments. Common themes among sites selling or promoting unproven treatments are: "Your follicles are blocked," or "It's not from the U.S., it's from Europe and China!" Other examples are given as well. The site also provides information on the Food and Drug Administration (FDA) and Federal Trade Commission (FTC) policy on scams as well as a link to the U.S. Consumer Gateway that allows for a search of products and companies that have been reported as fraudulent. The site also provides a "Clinical Crash Course" designed to help evaluate product test information and make sense of medical journal articles and abstracts. Suggestions for the questions to ask when reviewing the clinical literature are: who it was tested on, what areas of the scalp the product works on, what was measured, how long a period the product was tested, did the test have controls, how many people were studied, and were the study results published.

CONCLUSION

There are many legitimate sites providing quality health information on hair loss to be found on the Internet, some of which have been examined in this paper. However, there are many more sites that promote products that have no proven effectiveness. Statements implying that drug companies keep treatments secret or that the product being promoted has shown promising results outside the U.S. should make the user wary of the information on the site. Some sites even report clinical findings from reputable medical journals that on actual review of the journal articles do not report on that particular product, but just one of the ingredients in the product. Caution in judgment should be exercised when viewing commercial sites. The best advice is to encourage consumers to discuss the products they are considering using with a physician.

REFERENCES

1. Nolte, R. S., and Shaughnessy, A. F. "Medical Treatments for Balding in Men." *American Family Physician* 59 (8): 2189-96.

2. Sinclair, R. "Male Pattern Androgenetic Alopecia." *BMJ* 317(September 26, 1998): 865-9.

Index